PROTECTING BUILDING OCCUPANTS AND OPERATIONS FROM BIOLOGICAL AND CHEMICAL AIRBORNE THREATS

A FRAMEWORK FOR DECISION MAKING

Committee on Protecting Occupants of DOD Buildings
from Chemical and Biological Release

Board on Chemical Sciences and Technology

Board on Life Sciences

Division on Earth and Life Studies

NATIONAL RESEARCH COUNCIL
OF THE NATIONAL ACADEMIES

THE NATIONAL ACADEMIES PRESS
Washington, D.C.
www.nap.edu

THE NATIONAL ACADEMIES PRESS 500 Fifth Street, N.W. Washington, DC 20001

NOTICE: The project that is the subject of this report was approved by the Governing Board of the National Research Council, whose members are drawn from the councils of the National Academy of Sciences, the National Academy of Engineering, and the Institute of Medicine. The members of the committee responsible for the report were chosen for their special competences and with regard for appropriate balance.

This study was supported by the Defense Threat Reduction Agency (DTRA) under Award No. HDTRA1-06-C-0052. The content of this publication does not necessarily reflect the views or policies of DTRA, nor does mention of trade names, commercial products, or organizations imply endorsement by the U.S. government.

International Standard Book Number-13: 978-0-309-10955-0
International Standard Book Number-10: 0-309-10955-8

Additional copies of this report are available from the National Academies Press, 500 Fifth Street, N.W., Lockbox 285, Washington, DC 20055; (800) 624-6242 or (202) 334-3313 (in the Washington metropolitan area); Internet, http://www.nap.edu.

Printed in the United States of America.

THE NATIONAL ACADEMIES

Advisers to the Nation on Science, Engineering, and Medicine

The **National Academy of Sciences** is a private, nonprofit, self-perpetuating society of distinguished scholars engaged in scientific and engineering research, dedicated to the furtherance of science and technology and to their use for the general welfare. Upon the authority of the charter granted to it by the Congress in 1863, the Academy has a mandate that requires it to advise the federal government on scientific and technical matters. Dr. Ralph J. Cicerone is president of the National Academy of Sciences.

The **National Academy of Engineering** was established in 1964, under the charter of the National Academy of Sciences, as a parallel organization of outstanding engineers. It is autonomous in its administration and in the selection of its members, sharing with the National Academy of Sciences the responsibility for advising the federal government. The National Academy of Engineering also sponsors engineering programs aimed at meeting national needs, encourages education and research, and recognizes the superior achievements of engineers. Dr. Charles M. Vest is president of the National Academy of Engineering.

The **Institute of Medicine** was established in 1970 by the National Academy of Sciences to secure the services of eminent members of appropriate professions in the examination of policy matters pertaining to the health of the public. The Institute acts under the responsibility given to the National Academy of Sciences by its congressional charter to be an adviser to the federal government and, upon its own initiative, to identify issues of medical care, research, and education. Dr. Harvey V. Fineberg is president of the Institute of Medicine.

The **National Research Council** was organized by the National Academy of Sciences in 1916 to associate the broad community of science and technology with the Academy's purposes of furthering knowledge and advising the federal government. Functioning in accordance with general policies determined by the Academy, the Council has become the principal operating agency of both the National Academy of Sciences and the National Academy of Engineering in providing services to the government, the public, and the scientific and engineering communities. The Council is administered jointly by both Academies and the Institute of Medicine. Dr. Ralph J. Cicerone and Dr. Charles M. Vest are chair and vice chair, respectively, of the National Research Council.

www.national-academies.org

COMMITTEE ON PROTECTING OCCUPANTS OF DOD BUILDINGS FROM CHEMICAL AND BIOLOGICAL RELEASE

Co-chairs

DAVID R. FRANZ, Midwest Research Institute, Fredrick, Maryland
NORMAN L. JOHNSON, Referentia Systems, Honolulu, Hawaii

Members

WILLIAM P. BAHNFLETH, Pennsylvania State University, University Park
CYNTHIA BRUCKNER-LEA, Pacific Northwest National Laboratory, Richland, Washington
STEVEN B. BUCHSBAUM, Bill & Melinda Gates Foundation, Seattle, Washington
SHELDON K. FRIEDLANDER (deceased), University of California, Los Angeles
MURRAY HAMLET, U.S. Army Research Institute of Environmental Medicine, Kingston, Massachusetts
STUART L. KNOOP, Oudens Knoop Knoop + Sachs Architects, Chevy Chase, Maryland
ANDREW MAIER, Toxicology Excellence for Risk Assessment, Cincinnati, Ohio
R. PAUL SCHAUDIES, GenArraytion, Inc., Rockville, Maryland
RICHARD G. SEXTRO, E.O. Lawrence Berkeley National Laboratory, Berkeley, California
LINDA D. STETZENBACH, University of Nevada, Las Vegas
LINDA M. THOMAS-MOBLEY, Georgia Institute of Technology, Atlanta
DAVID R. WALT, Tufts University, Medford, Massachusetts

Staff

EVONNE P.Y. TANG, Study Director, Board on Life Sciences
ERICKA M. MCGOWAN, Associate Program Officer, Board on Chemical Sciences and Technology
DOROTHY ZOLANDZ, Director, Board on Chemical Sciences and Technology

BOARD ON CHEMICAL SCIENCES AND TECHNOLOGY

BOARD ON LIFE SCIENCES

Chair

Keith Yamamoto, University of California, San Francisco

Members

Ann M. Arvin, Stanford University School of Medicine, Stanford, California
Jeffrey L. Bennetzen, University of Georgia, Athens
Ruth Berkelman, Emory University, Atlanta, Georgia
Deborah Blum, University of Wisconsin, Madison
R. Alta Charo, University of Wisconsin, Madison
Jeffery L. Dangl, University of North Carolina, Chapel Hill
Paul R. Ehrlich, Stanford University, Stanford, California
Mark D. Fitzsimmons, John D. and Catherine T. MacArthur Foundation, Chicago, Illinois
Jo Handelsman, University of Wisconsin, Madison
Ed Harlow, Harvard Medical School, Boston, Massachusetts
Kenneth H. Keller, University of Minnesota, Minneapolis
Randall Murch, Virginia Polytechnic Institute and State University, Alexandria
Gregory A. Petsko, Brandeis University, Waltham, Massachusetts
Muriel E. Poston, Skidmore College, Saratoga Springs, New York
James Reichman, University of California, Santa Barbara
Marc T. Tessier-Lavigne, Genentech, Inc., South San Francisco, California
James Tiedje, Michigan State University, East Lansing
Terry L. Yates, University of New Mexico, Albuquerque

Staff

Frances E. Sharples, Director
Kerry A. Brenner, Senior Program Officer
Ann H. Reid, Senior Program Officer
Marilee K. Shelton-Davenport, Senior Program Officer
Evonne P.Y. Tang, Senior Program Officer
Robert T. Yuan, Senior Program Officer
Adam P. Fagen, Program Officer
Anna Farrar, Financial Associate
Tova G. Jacobovits, Senior Program Assistant
Mercury Fox, Program Assistant

Dedication

Dr. Sheldon Friedlander passed away on February 9, 2007, while he was serving on the Committee on Protecting Occupants of DOD Buildings from Chemical and Biological Release. The committee felt honored to have worked with him on this report. Sheldon was a world renowned expert in aerosol science and technology and a member of the National Academy of Engineering. His hard work and dedication to the National Academies and the field of aerosol research will always be remembered. Sheldon will remain fondly in the minds and hearts of all who knew him.

—The staff and committee members of the Committee on Protecting Occupants of DOD buildings from chemical and biological release

Preface

The Department of Defense (DOD) has identified acts of terror that employ biological or chemical airborne threat agents as a priority. Protecting buildings from release of biological and chemical airborne threat agents is only one aspect of DOD's effort to develop an active defensive program. In its simplest expression, protection of building occupants from biological and chemical airborne threats requires the creation and maintenance of a protective system sufficient to deter such an attack and to minimize its impact should an attack occur. The Immune Building Program was developed by the Defense Advanced Research Projects Agency for that purpose. As the Immune Building Program progressed from the research and development stage to the active deployment stage, DOD reassigned management of the program to the Defense Threat Reduction Agency (DTRA). Prior to the inheritance of that program, DTRA determined that a multifaceted look at building protection would be helpful in determining the future of building protection efforts within DTRA. The National Academies was asked to convene an expert committee to evaluate the proper terminology to exchange information; the metrics to be used to evaluate test beds and current deployments; the applicability of lessons learned from previous test beds and deployments—both in the military and the public domain; the protocols to be used; and the cost-benefit of different approaches and their relative risks. The ultimate goal of this study is to provide guidance in the complex-wide deployment of building protection to DTRA. Although the requirement is simply stated, its fulfillment is much more challenging.

The committee held four meetings in Washington, D.C., and St. Louis, Missouri, from September 18 to December 19, 2006. The committee was briefed by representatives of federal agencies and other entities that have deployed building protection or relevant programs. On-site visits of test beds and current deployments were made at Fort Leonard Wood, Missouri, and Washington, D.C. The committee also reviewed information available from the open literature, as well as new materials prepared by experts.

Early in the study, the committee attempted to provide a detailed implementation plan for the deployment and operation of building protection. As the committee delved more deeply into the study, it quickly became apparent that designing and implementing building protection is a complex process that involves many factors. Therefore, the committee's approach was to develop guiding principles to building protection. Although the charge concerned protection of military facilities, the guiding principles provided in this report are applicable to protection of public facilities as well. For many of the members of the committee, the challenges to provide defense from biological and chemical threats have been a lifetime concern, yet the present study provided an opportunity to examine a little-studied component of that defense.

We, co-chairs, wish to express our sincere appreciation to the National Academy project staff, who—behind the scenes—played an equal part with the committee in ensuring the quality of this report. We also want to express our personal appreciation to the individual members of the committee for the dedication and energy with which they tackled this challenging task. The report would not have been possible without the perspectives of these experts, their valuable time commitment, and their patience in integrating our diverse disciplines.

David R. Franz
Norman L. Johnson
Co-chairs, Committee on Protecting Occupants of DOD Buildings from Chemical and Biological Release

Acknowledgments

This report is a product of the cooperation and contributions of many people. The members of the committee thank all of the speakers who briefed them on different programs. (Appendix C contains a list of presentations to the committee.)

This report has been reviewed in draft form by persons chosen for their diverse perspectives and technical expertise in accordance with procedures approved by the National Research Council's Report Review Committee. The purpose of this independent review is to provide candid and critical comments that will assist the institution in making its published report as sound as possible and to ensure that the report meets institutional standards of objectivity, evidence, and responsiveness to the study charge. The review comments and draft manuscript remain confidential to protect the integrity of the deliberative process. We wish to thank the following for their review of this report:

Daniel Cousins, Massachusetts Institute of Technology Lincoln Laboratory
Charles Haas, Drexel University
Charles Kolb, Aerodyne Research, Inc.
Benson Kwong, Project Management Services, Inc.
Lewis S. Nelson, New York University School of Medicine
Leslie Robertson, Leslie E. Robertson and Associates, R.L.L.P.
Scott Rusk, Biosecurity Research Institute
Timothy Swager, Massachusetts Institute of Technology
James Woods, HP-Woods Research Institute

Although the reviewers listed above provided constructive comments and suggestions, they were not asked to endorse the conclusions or recommendations, nor did they see the final draft of the report before its release. The review of this report was overseen by Dr. David C. Bonner and Mrs. Hyla S. Napadensky. Appointed by the National Research Council, Dr. Bonner and Mrs. Napadensky were responsible for making certain that an independent examination of this report was carried out in accordance with institutional procedures and that all review comments were carefully considered. Responsibility for the final content of this report rests entirely with the authoring committee and the institution.

Contents

Summary

Several widely publicized attacks using biological and chemical threat agents in the last two decades have increased the urgency of protecting buildings and the materials, persons, and critical operations housed in them from these threats. To address that need, the Defense Advanced Research Projects Agency (DARPA) initiated the Immune Building Program to design, implement, and test a building protection system to make military buildings and their occupants less attractive targets for attack with biological or chemical threat agents. The Defense Threat Reduction Agency (DTRA), which is scheduled to assume responsibility for test beds[1] and other results developed by DARPA's Immune Building Program, asked the National Academies to convene a committee to consider existing work on preventing and mitigating the effects of airborne biological or chemical threat agents released within or infiltrated into built structures. The committee was asked to provide general principles that can be derived from those studies and existing test beds and to discuss the cost, benefit, and risks of potential protection schemes (see Appendix A for the complete Statement of Task). It is hoped that the results of this study will provide guidance for future investments in the Immune Building Program and other building protection efforts. The study committee included experts in technologies related to aerosols, biological and chemical warfare threats, detection and identification of biological and chemical threat agents, medical countermeasures, building design and operations, indoor airflow, and risk assessment (see Appendix B for committee member biographies). To gather information to address its task, the committee

[1]An environment created for testing that contains the integral hardware, instrumentation, simulators, software tools, and other support elements to approximate real-world situations.

heard about the purpose, feasibility, and capability of prior and current building protection schemes from representatives of various agencies and contractors involved in building protection programs and test beds. Based on these briefings, other documents, committee deliberations, and the committee's collective expertise, this report highlights basic principles and lays out the variables and options to consider in designing and implementing building protection against biological and chemical threats.

FACTORS THAT INFLUENCE THE DESIGN AND IMPLEMENTATION OF BUILDING PROTECTION

Appropriate design and implementation of building protection is determined by multiple factors, including (1) threat types; (2) activities housed within the building and the mission of those activities; (3) the level of protection sought; (4) limitations posed by procurement type, quality of design and construction, maintenance, and wear and tear of the building; and (5) availability of resources.

Threat Types and Threat Agents

This report addresses two threat types: airborne releases of biological threat agents and airborne releases of chemical threat agents. "Threat agent" refers to the biological or chemical agent used in an attack. Biological threat agents considered here include bacteria (vegetative and spores), viruses, and products of organisms (toxins). Chemical agents considered here include those that can be dispersed as droplets or vapors. Because of the wide variety of biological and chemical agents, their symptomatic progressions, and associated fatality rates, the committee concluded that the best classification of threat agents is according to the two most critical properties related to current vulnerabilities in building protection—the ease of timely detection and the ease of timely treatment (see Figure S-1). This classification treats biological and chemical threats equally and addresses vulnerabilities from unknown threats. The ability to detect agents ranges from agents that are visible to the naked eye to those that are difficult or impossible to detect at levels of concern even with sophisticated equipment (see Figure S-1). The time window for treatment likewise varies from several days for threat agents that have long incubation or latent periods (slow-acting) to minutes for threat agents that produce rapid onset of illness or even death (fast-acting). Protection from difficult-to-detect and fast-acting threat agents is obviously the most challenging building protection scenario.

Activities and Mission

Buildings are constructed for different purposes and have different activities occurring within them. The need for protection varies among structures based

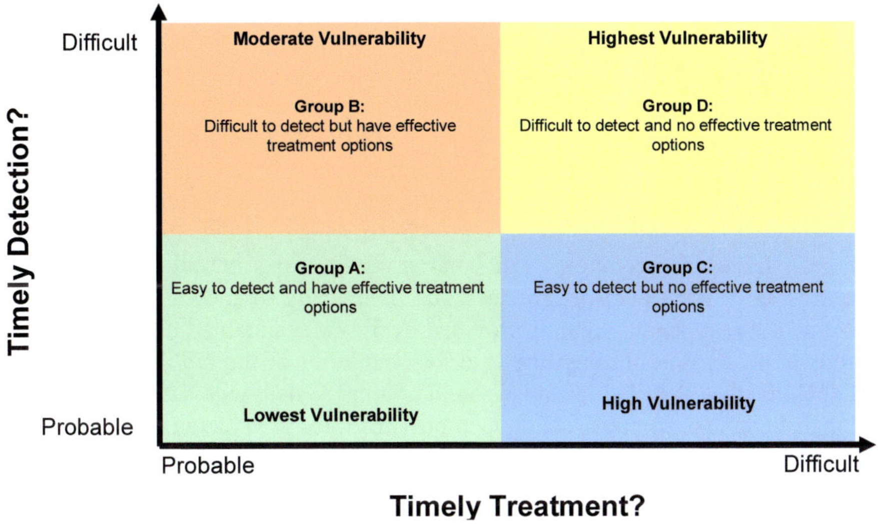

FIGURE S-1 Grouping threats on the basis of the ability to detect the threat and respond in a timely manner.

on the activities and missions they serve. For example, a structure used as a warehouse has different protection needs from one used to house troops or from one used as a critical operations control center or a hospital. For some buildings, a shut down of several days or several weeks for decontamination and recommissioning after a biological or chemical attack might be acceptable; in other structures, continuity of operations is paramount. Building use, with respect to its contents or occupants and their activities and mission, should play a large role in determining the type and level of protection needed.

Levels of Protection

Different strategies can be chosen to provide varying levels of protection across the spectrum, from no protection to strategies designed to totally eliminate the exposure of personnel to an agent. Active or passive strategies can be used to protect against threat agents. Passive measures provide protection by using approaches that do not include identification and detection technologies. For example, compartmentalization of spaces within the building, continuous cleansing of airstreams, visual recognition of threats and their effects, and relying on the integrity of the building as a whole to protect building occupants and contents are passive approaches to protection. Conversely, active measures use identification and detection technologies to recognize the presence of a threat agent and trigger a response, but as a result of technological and operational complexity, they have more complex operation and higher risk of failure (including false positive and false negative rates) than passive strategies. The committee defined four levels of protection (LPs) that are not absolute but can be used to illustrate the components needed and options available for achieving desired protection goals. These levels of protection are qualitative, like the biosafety levels of microbiological and biomedical laboratories. In some circumstances, advanced protection options could be implemented without some of the low-level components.

- Level of protection 1 (LP-1) is a low-level passive protection that has no sensors or additional options installed specifically to address biological and chemical threat agents. LP-1 is provided by a well-maintained building that has minimal air leakage through the exterior or interior of the building and has an HVAC (heating, ventilating, and air-conditioning) system with sufficient filtration and air exchange. Its construction methods are aimed at reducing particulates and chemical vapors in the finished structure.
- Level of protection 2 (LP-2) is a high-level passive protection that does not utilize sensors. Options for achieving LP-2 include site selection, addition and upgrade of filters and adsorption units specific to biological and chemical threats, compartmentalization and overpressurization of building interiors, filtration of outdoor air, relocation of outdoor air intake vents, local air-washes,[2] security protection in the surrounding area, and appropriate operational responses.
- Level of protection 3 (LP-3) is a low-level active protection designed to detect and identify threat agents in time to execute therapeutic responses, but not quickly enough to warn occupants of the threat before exposure occurs. LP-3 requires a broad-spectrum detection and identification system that determines a threat agent within a time period necessary for operational response and treatment.
- Level of protection 4 (LP-4) is a high-level active protection that can detect and identify a threat agent in time to mitigate the release. LP-4 can detect a threat early enough to make operational responses that prevent exposure, such as redirecting ventilation or donning personal protective equipment.

The difference between LP-1 and LP-2 is the inclusion of options specifically for protection from biological and chemical airborne threats in LP-2. The presence of detection and identification technologies for biological and chemical airborne threats defines LP-3 and LP-4. Automated response to threat detection separates LP-3 from LP-4. In general, active protection (LP-3 and LP-4) has fewer vulnerabilities when implemented in conjunction with passive protection. In this case, LP-3 or LP-4 is likely to include the virtues of LP-1 and LP-2.

---

[2]Local air-washes are areas that are provided with isolated, enhanced laminar airflow with local filtration at the returns.

Procurement Method and Building Type and Condition

Building procurement method and building type could limit the options that are available and the subsequent level of protection achieved in a building. Leased buildings that are partially or completely occupied by federal tenants likely offer security planners fewer options than federally owned buildings because of contractual constraints. Building type or interior layout affects a building's vulnerability to threat agents because compartmentalization or the lack thereof can affect the spread of threat agents throughout the building.

A protection system designed to be integrated into newly constructed buildings is likely to have fewer limitations than a retrofit to an existing building. Building protection systems cannot be standardized or even generalized because the physical characteristics of buildings—their age, quality of construction, "leakiness"—and ongoing activities inside vary greatly within and across military and civilian sectors.

Financial Resources

An integral part of the decision-making process when implementing a building protection system is cost consideration. The budget for building protection obviously limits the design and implementation of the protection system. Protection systems in newly constructed buildings could be more or less expensive than retrofitting an existing building depending on protection goals. In general, the former is less expensive than the latter given the same level of protection, especially if security needs are anticipated early in the pre-design and design phases and are identified in a threat and risk assessment.

All monetary costs associated with a protection system within its lifetime (life-cycle costs) need to be considered prior to its implementation. Fitting a building with protection from biological and chemical airborne threats would be unwise if a budget for operations and maintenance costs cannot be ensured. Inadequate long-term operation and maintenance budgets can defeat the performance objectives of the building and render investments in building protection worthless. Complete life-cycle costs of a building protection system include the initial costs of planning, design, and construction; cost of purchase, installation, and periodic and preventive maintenance; cost of operation, repair, and replacement of parts; and cost of upgrade of all its components. It should be noted that in government facilities, long-term budgeting and planning for costs of operation and maintenance are the exception rather than the rule.

Goals and Objectives for Protection

The goals and objectives of building protection vary depending on the mission and activities of each building. Clear definitions of goals and objectives for building protection prior to designing, implementing, or deploying a protection system are essential so that appropriate components of the system and metrics for evaluation can be chosen. Because of variations in goals and in the factors that influence the feasibility of building protection, protection systems clearly cannot be designed generically. In defining goals and objectives, factors that influence building protection can provide guidance in determining the feasibility and limitations of desired protection options, thus determining the level of protection that can be achieved. The committee has developed three recommendations related to the design planning for building protection.

Recommendation 1: Clear and realistic building protection goals and objectives should be defined prior to deploying protection systems.

Recommendation 2: Building protection systems should be designed and implemented on a case-by-case basis for each structure to be protected.

Recommendation 3: Life-cycle costs should be planned for prior to deploying building protection systems.

COMPONENTS FOR BUILDING PROTECTION

A number of components can be used in building protection and applied in different combinations to achieve different levels of protection. Components that can be designed, modified, installed, or implemented to enhance building protection include the following:

- *Building design and planning strategies.* These are passive strategies for enhanced physical security, such as choosing a site with adequate standoff from neighbors and protecting sensitive areas (such as fan rooms and filtration and pump rooms) from unauthorized access.
- *Heating, ventilating, and air-conditioning systems.* HVAC can be used as a passive strategy that enhances protection from biological and chemical airborne threats through zoning, enhanced particulate filtration, the addition of continuous gas and vapor protection, and sensing and active control of airflows and air treatment devices. HVAC can also be used as an active strategy if it is coupled with validated sensor networks and used for threat mitigation.
- *Filtration.* Using particulate and vapor adsorption filters with the highest feasible efficiency and replacing filters routinely are passive strategies to enhance protection for even those threat agents that cannot be detected.
- *Detection and identification technologies.* These technologies are a part of an active protection strategy. They can include a sampling system to identify the threat agent; sensors that actively monitor the presence of threat agents, either periodically or continuously; and triggers that measure an event and initiate

another action. The use of these technologies singly or in combination could enhance the ability to respond to a threat by exposure prevention, mitigation, or treatment.

- *Operational responses.* An overall concept of operations is an integral and necessary part of planning and preparation because inappropriate response actions can increase the hazard to occupants or missions. Operational responses could include active HVAC responses, shelter-in-place,[3] use of personal protective equipment, or evacuation. Developing and practicing operational responses will maximize protection from biological and chemical threats.

A protection system can be designed to use passive or active approaches or both. Although active approaches, detection, and identification technologies are necessary to achieve LP-3 and LP-4, the components of passive protection (LP-1 and LP-2) are integral in these active systems. Sensor systems cannot perform to their best capacity without the high air quality provided by LP-1 and LP-2 systems. Similarly, LP-3 and LP-4 systems are not useful if operational plans are not available and in place to respond to alarms.

ROLE OF TEST BEDS AND EXISTING DEPLOYMENT

The defense community has developed test beds and field studies to evaluate the use of protection components, and it has active deployments that can provide further data. Because the factors that influence protection and protective components and technology vary from building to building and change over time, the ability to extrapolate results from test beds and deployments is limited. However, test beds, field studies, and deployments are valuable for evaluating the performance of individual sensors under realistic conditions and for evaluating the performance of integrated sensor systems and building protection systems (with or without sensors). One advantage of test beds is that they can be configured and challenged in ways not possible in an operational facility. Data collected on degradation, maintenance, and operational and life-cycle costs of building protection systems in test beds—and, in time, from operational buildings—can be used as points of reference for future analyses.

Because test facilities cannot faithfully duplicate specific operational buildings and data from actual biological or chemical attacks are sparse, modeling and simulation are necessary for assessing building protection. Data from test beds and existing deployments are important for developing and refining models, but uniform test methods for data collection in the test beds and deployments are necessary for data comparison.

---

[3]Shelter-in-place means taking refuge in a small, interior room with no or few windows. See the following website for more information: *http://www.redcross.org/services/disaster/beprepared/shelterinplace.html.*

METRICS AND SYSTEM EVALUATION

A well-conceived building protection strategy includes metrics to measure the success of that strategy against building protection objectives. In addition to measuring the performance of a protection system, metrics can provide a common basis for comparison between different deployments and demonstrations. Protection metrics and operational performance metrics are used to gauge whether a building protection system is performing as planned. Protection metrics, which can include fraction of building exposed, fraction of occupants exposed, and lives saved, usually measure against the protection goals. Evaluation of these performance metrics requires either comprehensive testing in the facility or use of modeling tools to infer values for the desired metrics. Operational performance metrics, which may be less quantifiable than protection metrics, include response time, user acceptance, adaptability to new technologies, and management of disruptions caused by false alarms.

There is no universal set of metrics that can be used to assess protection systems of all buildings because of the uniqueness of each building, its use, and the goals and objectives of its protection. The goals and objectives shape the design and implementation of the system, as well as the appropriate metrics to measure system performance. The committee makes the following recommendation related to metrics:

Recommendation 4: Because goals and objectives for protection drive the choice of building protection system for each installation, metrics for a building protection system should be based on these same well-understood, clear goals and objectives.

A FRAMEWORK FOR DECISION MAKING

A systematic process that takes into account the building's vulnerabilities and risks of attacks, its physical limitations, the budget, and options for protection using risk assessment and management approaches is needed to guide decision making and cost-benefit analysis for building protection. Because of the complex interactions of factors that determine the design of a building protection system, establishing a systematic process that weighs these factors against different protection options would help DTRA and other agencies to design appropriate protection systems. Such a process would help optimize protection within the physical limitations of the building, technological limitations of the components (such as the HVAC system, filters, and sensors), and financial constraints.

Recommendation 5: Prior to implementation of a building protection program, the Department of Defense (DOD) should establish a complete framework for building protection that guides decision making for each building to be protected. The decision-making framework should

consider the following steps: (1) defining the objectives of building protection; (2) preparing a threat assessment; (3) establishing a risk assessment; (4) developing a case-by-case plan for building protection; (5) conducting a risk management analysis; and (6) analyzing costs and benefits using appropriate metrics and modeling and simulation tools as needed. The complexity of steps in the framework and the time required for each step will depend upon the program and building protection objectives.

PLAN FOR THE FUTURE

Although buildings, threats, and vulnerabilities are unique and dynamic, there are some guiding principles for designing and implementing building protection. The level of protection needed depends on the goals and objectives defined, the vulnerabilities of the building, and the risks of attack, which could change over time as the building ages and the threat spectrum evolves. New threat types could be developed and deployed as scientific advances remove technical barriers. Cutting-edge building materials and techniques are being developed not only to provide basic protection but also to provide a "healthy" environment for building occupants. Predicting future needs and capabilities will be difficult, but designing today with likely future capabilities in mind can lower life-cycle costs of building protection systems and facilitate the utilization of future options.

Recommendation 6: Building protection should be designed to accommodate changing building conditions, emerging threats, and changing technology. Both the deployed building protection and the framework for deploying the building protection (proposed in Recommendation 5) should be reviewed periodically for sustained performance in light of changing resources and threats.

Protecting buildings from biological and chemical airborne threats is a complex matter subject to many variables. These variables have a striking impact on the feasibility and capability of the desired protection system. A well-defined strategy for protection, starting at the design phase and continuing through the deployment phase, combined with sound decision making, can lead to the best options for reaching building protection goals now and into the future. Although the principles of building protection might not change substantially over time, the technologies for protection and the threats will likely change; therefore, periodic reviews of strategies for building protection might be necessary.

1

Introduction

In the age of asymmetric warfare, buildings and other enclosed spaces have become potential targets of biological or chemical attack. In some ways, humans inside buildings are even more vulnerable to terrorist attack than humans outside. The anthrax incidents of 2001 demonstrated the impact of such attacks on not only the occupants but also the operations conducted inside the building. The Curseen-Morris postal facility (formerly Brentwood) was not in operation for more than two years as a result of contamination with *Bacillus anthracis* spores. The critical national security mission of the Department of Defense (DOD) demands healthy employees working in functional buildings; disruption of operations or prolonged closure is unacceptable. Therefore, DOD, the Department of Homeland Security (DHS), and other agencies have programs and test beds[1] to study and improve the means of protecting buildings from biological and chemical attacks. These test beds and programs include the Defense Advanced Research Projects Agency's (DARPA's) Immune Building Program (Bryden, 2006), the Joint Program Executive Office for Chemical and Biological Defense's Guardian Program, and the Safe Building demonstration in Salt Lake City. What are the general principles learned from those programs and test beds? What metrics could be used to assess system performance? How do existing programs and test beds inform the design and implementation of protection systems for other buildings? The Defense Threat Reduction Agency (DTRA) commissioned the National Research Council (NRC) to convene a committee to draw together lessons learned and principles that might inform the design and

[1]An environment created for testing that contains the integral hardware, instrumentation, simulators, software tools, and other support elements to approximate real-world situations.

implementation of protection systems against biological and chemical airborne threats for new and existing DOD buildings.

STUDY CHARGE AND SCOPE

The committee was charged to address the following issues:

- What metrics of performance are relevant to evaluate existing studies and to use existing facilities as effective test beds for validating tools, testing systems, and facilitating system technology transfer? Where a metric is not relevant to all situations, identify and discuss its appropriate application. Discuss situational use of a combination of all relevant metrics where appropriate.
- What terms and definitions are required—for example, Tier 1 detector, trigger, high-impact response, confirmatory test, and so forth—to allow communication and comparison among programs?
- Consider the current protocols and setup of existing systems in use, including both DOD and non-DOD efforts. What are the general features of existing test bed facilities? Are there significant features in common? Do existing facilities differ in significant ways, and how can these differences be exploited to forward our understanding of building protection?
- What collective principles can be derived from current building protection efforts? How can information gained from a test bed facility be extrapolated to operational buildings with completely different designs?
- What is the cost-benefit of internal building monitoring? Suggest a tiered approach with varying levels of detection and protection capability, comparing the relative cost-benefit among the tiers. Perform this assessment for both new building construction and building retrofit, and correlate to an appropriate metric (lives saved or fraction of the building exposed).
- Compare and discuss the relative risks of the possible tiers in a tiered approach to chemical and biological protection efforts, from a baseline of no protection efforts up to and including a fully protected building. Consider risks associated with building retrofitting, extrapolating test data to buildings differing from test bed buildings, possible system degradation over time, et cetera.

An evaluation of the performance of building protection technologies or existing protection systems and test beds was not the intent of the study. Rather, it aims to provide DTRA guidance on investment, design, and implementation of building protection. The scope of this study is limited to airborne biological and chemical threat agents, even though explosives and radiological threat agents could be used in terrorist attacks as well. Although the goal is to protect occupants and operations within a building, biological and chemical airborne threats inside or outside a building could affect the occupants and activities within. Therefore, this report covers both inside and outside releases that might affect building occupants and operations. The purpose of building protection is ultimately to collectively protect humans and to allow continuation of their activities in the building with minimal disruption. Most approaches to protection of occupants also protect contents; therefore, the committee focused primarily on protection of occupants, and thus operations. The protection of service members on the battlefield was not the focus of this study, but the committee recognizes that "the battlefield" is different now from what it was during the Cold War. Formerly, engagement in conventional warfare was a well-defined confrontation of opposing parties where the target was the opposing army. Alternatively, asymmetric warfare often occurs between parties of unequal power where the opponent's vulnerabilities are attacked using unconventional weapons (for example, biological, chemical, nuclear, radiological) and often targeting anyone, anywhere, at any time—including the civilian population. In this age of asymmetric threats and global terrorism, the nation's military installations at home and abroad might be more likely targets of biological and chemical attacks than troops on the traditional battlefield.

COMMITTEE'S APPROACH TO ADDRESSING THE CHARGE

Building protection is a complex and dynamic issue that depends on the requirements regarding protection and the threat types that the system is intended to protect against. The design of an appropriate protection system depends on the goals and objectives of building protection and the threat types that it is intended to protect from. Chapter 2 defines the threat types and reviews biological and chemical threat agents that might be used in an attack. The committee grouped threat agents on the basis of the capability to detect them and to provide treatment for exposed victims. The chapter also discusses the factors that shape the goals and objectives of a building protection system.

To facilitate the discussion of protective capability in Chapter 3, the committee defines four levels of protection (LPs)—LP-1, low-level passive; LP-2, high-level passive; LP-3, low-level active; and LP-4, high-level active—that can be achieved considering performance metrics, protection objectives and goals, and the proposed threat. A protection system involves many components, some of which are included in the building design (for example, the heating, ventilating, and air-conditioning [HVAC] system). Others are installations designed specifically for protection from biological and chemical airborne threats (for example, sensors). Chapter 3 covers options for building protection, including levels of protection and components that can be used to achieve preset building protection goals and objectives.

Chapter 4 discusses metrics and evaluation criteria that could be used to assess building protection. Chapter 5 describes prior and existing programs and test beds in building protection that the committee considered.

Designing and implementing an appropriate building protection system depends on the interaction of many factors, including budget, objectives of protection, and activities in the facility. Chapter 6 presents a structured approach to the design and deployment of a building protection system that is based on risk assessment and management. That chapter also proposes an analytic process that is based on currently available and proven methodologies to assess costs and benefits of a building protection system. The committee presents its conclusions and recommendations in Chapter 7.

GOALS AND ANTICIPATED IMPACT OF THE STUDY

As in any emergent situation, there is a window of operational influence following a biological or chemical attack on a building. The time during which one can make critical decisions or take actions of low or high regret may be extremely short (minutes).[2] Preparation—technical and operational—can be advantageous to protecting building occupants, but at a cost. It is the committee's hope that this report will assist DOD in making cost-benefit decisions that will maximize the window of operational influence, save lives, and maintain operational facilities at an acceptable cost. Although this study was commissioned by DTRA for the protection of military buildings, the committee's findings and recommendations can, obviously, also be applied to nonmilitary buildings.

[2]Low-regret or high-regret actions refer to responses to a situation that could incur a low or a high degree of remorse. Whether a certain action is a low-regret or a high-regret response is a value judgment.

2

Factors That Influence Building Protection

Buildings do not merely provide shelter for and enable the performance of their occupants; they are built to resist damage from fire, earthquakes, wind, and other hazards. Safety measures in buildings reduce the probability of injury to occupants and thereby minimize the disruption of operations in the facilities. However, threats to buildings are not limited to natural and industrial disasters and unintentional releases of hazardous substances. There have been intentional attacks on buildings in the United States in recent years, such as the attacks on the Alfred P. Murrah Federal Building in Oklahoma City and on the World Trade Center in New York City, and the anthrax incidents in various locations. Although the motive behind the mailing of letters containing *Bacillus anthracis* in 2001 is not known, the outcome was essentially an attack on humans in buildings, which both harmed the victims and disrupted operations within the buildings.

To protect against such attacks, it is important to consider various types of threats and their potential modes of delivery. Before designing any protection system for a building, the goals of protection have to be established. The goals are often driven by the mission of and activities in the building. The ability to meet these goals is partly constrained by the building type and procurement and the budget for the protection system. The Department of Defense (DOD) defines a DOD building as any building or portion of a building owned, leased, privatized, or otherwise occupied, managed, or controlled by or for the department (DOD, 2003). In other words, DOD buildings vary in type, procurement, and mission, all of which influence the design and implementation of building protection from biological and chemical airborne threats. This chapter discusses the factors that determine the goals and objectives of building protection. Cost, a consideration in the design of a protection system, is also a criterion for evaluation and is discussed in detail in Chapters 4 and 6.

THREAT TYPES AND AGENTS

"Threat type" refers to different categories of methods and agents that a perpetrator could use in an attack—biological, chemical, and radiological agents and explosives in combination with a dissemination means. "Threat agent" refers to the specific biological or chemical agent used in an attack, such as *B. anthracis* or sarin. Within the two threat types that are considered in this report, biological and chemical, many threat agents could be used against buildings and occupants. Different threat agents can have different physical properties—they could be solid particulates, liquid, or vapor. There are advantages in choosing different threat agents or physical properties depending on the configuration, location, and condition of the specific target building and the emergency preparedness of the building and its occupants. Target selection depends on building design; security in place; design of the HVAC system; personnel schedules; availability of agents; and technical capability of the perpetrator(s). Agent selection considerations might include availability, ease of handling, volume required, and the intent and skill of the perpetrator. Agent-disseminating devices can be simple or complicated; volatile chemicals or very fine prepared powders can be disseminated more easily than nonvolatile liquids. Taking all of the variables into account, it is clear that no building can be completely protected from a sophisticated, well-trained, determined aggressor. In buildings that are at high risk of a terrorist attack, the goal of building protection is to reduce vulnerabilities to and consequences of external or internal delivery of threat agents into buildings. Enhanced physical and operational security is as important as sensors and mechanical response systems in reducing the likelihood and impact of an attack.

Biological Threat Agents

Biological threat agents include bacteria (vegetative and spores), viruses, and products of such organisms as toxins. They can be delivered as nonvolatile particulates or suspended liquids. The most likely mode is the delivery of aerosolized particles in the size range of 0.3 to 10 μm. Generally, particles larger than 10 μm are less likely to reach the lungs, but they could be lodged in the mucosa of the nasal passage or the pharynx or simply settle out of the airstream and contaminate the area. Biological threat agents include disease-causing microorganisms and protein or low–molecular weight toxins. Biological threat agents are often mixed with inert materials that enhance dispersal and improve stability (see NRC, 2005a, Table 2-1). Once detected, microorganisms and toxins can be identified definitively by various means of analysis.

Microorganisms

Disease-causing organisms that are most likely to be used as threat agents include bacteria in vegetative or spore form and viruses. These threat agents can be persistent—that is, they can remain infective for a relatively long time—or nonpersistent depending on their innate properties, environmental conditions, and whether they have been processed to enhance their stability. They can be contagious (easily transmissible from person to person, especially by the aerosol route without physical contact) or noncontagious (poorly transmissible, typically by direct contact with the agent). The ease of spread depends on the organism's ability to survive different environmental conditions; its tropism (the involuntary response of an organism or part of an organism toward or away from external stimuli) and presence in body fluids (such as airway secretions); its infectivity (number of organisms needed to initiate infection), pathogenicity, and incubation period; a suitable host population; and other factors.

Noncontagious threat agents have many of the characteristics of contagious ones, except that noncontagious organisms infect individuals only by direct exposure and cannot be transmitted from one person to another. Thus, individuals exposed to noncontagious threat agents are limited to those who come in direct contact with the initially released material or with material that remains viable in the environment after the primary aerosol cloud has dissipated.

Toxins

Many of the most lethal toxins, such as botulinum, are less toxic if inhaled (by the aerosol route) rather than injected into the body. However, inhaled doses of certain toxins can lead to illness or death within minutes to hours. Toxins would typically be distributed as particulate aerosols and are not transmissible. Unlike volatile chemicals, some toxins are persistent and could require active mitigation, although many toxins are less stable than human-made chemicals.

Impact of Biological Threat Agents

Infectious biological threat agents pose a serious threat because of the relatively small amount of material (milligram to gram quantities) required to infect many building occupants. Biological threat agents are typically odorless and not visible to the naked eye as aerosols so that humans cannot detect them during exposure. Some, such as the spores of *B. anthracis*, are stable and significantly complicate the ability to resume operations (NRC, 2005a). Infectious biological threat agents that are transmissible complicate response and recovery because the impact of their release can reach beyond the building attacked. Therefore, rapid identification of the agent used is important to minimize its spread and transmission. Methods for delivery of aerosolized biological threat agents are more limited that those of chemical threat agents. Whereas chemical threat agents can be dispersed by explosive devices, many biological threat agents are highly flammable and sensitive to heat, both of which restrict the methods of delivery.

Chemical Threat Agents

Chemical threat agents can be in the form of particulates (solid particles or liquid aerosols) or in the gaseous state (gases and vapors). Chemical threat agents include traditional military blister and nerve agents and toxic industrial chemicals (TICs) and materials (TIMs). TICs and TIMs are common industrial chemicals manufactured, stored, and transported in large quantities, and the former has an appreciable (undefined) vapor pressure at 20°C. The industrial chemicals are broadly available worldwide, although they are often less acutely toxic than chemical warfare agents.

Blister and Nerve Agents

Blister agents, also called "mustard" agents, are considered less of a threat in the twenty-first century because they have not been used since World War I and their manufacture as weapons of war is probably very limited today. Blister agents include formulations of sulfur mustard (H, HD, and HT) and arsenicals such as Lewisite. Nerve agents include VX (*O*-ethyl *S*-[2-diisopropylamino]ethyl]methylphosphonothiolate), GB (sarin), and GD (soman). The Tokyo subway attacks of 1995 involved the use of the poor-quality, homemade nerve agent sarin. Nerve agents are organophosphonate compounds that are extremely toxic at very low doses and can cause convulsions and death within minutes. Because of their potency, nerve agents are likely to be used in small quantities by perpetrators and would pose a threat if they were released inside a building.

TICs and TIMs

TICs and TIMs are produced in large quantity as chemical feedstocks, and products are typically transported to their destinations in liquid form via railcar or tanker trucks. Thousands to millions of gallons of these substances could be present in a shipment. Although they might have lower acute toxicity than blister and nerve agents, they are pervasive and available in large quantities (DOJ, 2000). TICs and TIMs are hazardous because they are toxic by definition and because many of the products, intermediates, and by-products are highly volatile and can form vapor clouds upon release. Some TICs are also flammable. A large spill or an intentional introduction from outside a building could result in rapid distribution of large quantities of the agent within a building through the HVAC system or infiltration. TIC plumes that are heavier than air tend to be persistent because they gravitate toward the ground and are unlikely to disperse quickly.

This is potentially problematic in buildings where air intakes are not elevated, increasing the possibility that the contents of the plume could infiltrate the building. However, such gases might not distribute readily.

In addition to their toxicity, some TICs and TIMs pose a health threat because they can cause respiratory, skin, and eye irritation, but they might not be easily distributed. Although many TICs do not pose an immediate hazard to life, exposure in sufficient amounts can cause fatalities in most cases. A comprehensive threat assessment (see Chapter 6) would evaluate the spectrum of TICs that might be of concern and their toxicological properties. Exposure to TICs generally causes such respiratory symptoms as coughing and respiratory distress up to and including pulmonary edema, and it induces a variety of systemic effects. Examples of TICs include chlorine, bromine, hydrochloric acid, anhydrous ammonia, and organics such as benzene, toluene, and other industrial solvents.

Impact of Chemical Threat Agents

The vapor pressures and volatilities of the various chemical threat agents are particularly relevant to their detection as airborne vapors. The volatility of sarin is comparable to that of water or volatile organic compounds such as limonene and cyclohexanone (24,000 to 28,000 mg m^{-3}). In contrast, VX has a low volatility, on the order of long-chain aliphatic waxes (for example, the volatility of docosane [$C_{22}H_{46}$] is about 17 mg m^{-3}). TICs of concern as threat agents typically have higher vapor pressures than many other chemical threat agents; they could be attractive to attackers because they can be dispersed more easily than compounds with low vapor pressure.

A variety of chemical threat agents such as mustard, blister agents, nerve agents, and acute toxic materials (such as cyanide gas) can be released as vapors. Some agents are toxic at extremely low concentrations and pose an immediate threat to the lives of exposed individuals. The time between exposure and death is usually only a few minutes. Others have longer-term impact on health. Delivery of volatile agents can be as simple as opening a vial and allowing material to vaporize into the target area. Less-volatile materials are more effectively disseminated as aerosols. The degrees to which chemical agents have good warning properties (distinctive odors or minor irritation at concentrations lower than those that induce severe acute toxicity) vary. Threat agents with good warning properties are readily detectable, but because of variability in the properties of threat agents, a unique threat assessment is necessary for each building. The assessment is an important aspect of having a systematic approach for building protection as discussed in later chapters.

Once a threat agent is detected, the best way to avoid widespread exposure is to evacuate the contaminated area and allow the material to evaporate while increasing air circulation and exhausting air from that isolated area. From the perspective of building protection, chemical threat agents with relatively high vapor pressure pose only a short-term threat to the building because they are cleared by evaporation and dilution with uncontaminated outdoor air. In general, chemical attacks require larger quantities of material to affect a given volume of air space within a building than do biological threat agents. Therefore, keeping certain illicit chemical threat agents out of a building through physical and operations security might be easier than keeping out biological threat agents.

Delivery of Threat Agents

There are numerous methods through which threat agents could be delivered to the inside of a building to harm humans and disrupt operations. The two fundamental approaches for agent dissemination are through outdoor release (leading to building air contamination through infiltration or uptake through HVAC air intakes) and through release directly within the building. Attack from the outside might be done through a broad release upwind from the facility or a focused release near an air intake or other access point. Inside attack could be through an internal release by a visitor, delivery person, or trusted employee or through a delayed release from munitions or a delivery system sent through the mail or left behind. A threat agent release could cause disruption through rapid or delayed onset of illness or rapid or delayed death with or without long-term contamination of the facility.

Classification of Threats

Because there are many combinations and variations of threat agents, delivery types, and targets, it might be useful to group threats on the basis of the ability to detect the presence of an agent and the ability to treat exposed individuals. Such grouping provides a structure for managing threats. The committee proposes four groups of threats that are based on the ability to detect and treat them (Figure 2-1). The term "detect" refers to detection of a threat agent by visual identification or a sensor system, by a symptomatic response, or by clinical determination. The term "treat" refers to treatment of exposed victims. The proposed groups account for both agent characteristics and available capabilities so that they are still applicable even if the threat agents or capabilities of delivery change or advance over time. By analyzing a given building, mission, and situation in the context of the proposed scheme, one might be able to arrive at a relative cost estimate and highlight vulnerabilities that might not be otherwise obvious. The four groups (A–D) generally represent situations that proceed from less vulnerable (A) to more vulnerable (D).

- **Group A—Can Detect, Can Treat.** Examples that apply to this group include threats that are visible (for example, large quantities of white powder such as that seen in the Hart Senate Building), that cause treatable disease al-

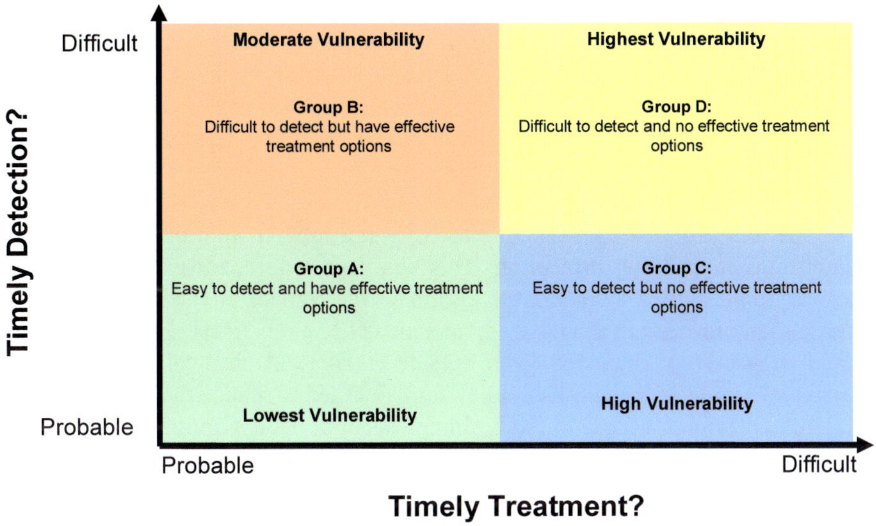

FIGURE 2-1 Grouping threats on the basis of the ability to detect the threat and treat it in a timely manner.

most immediately (for example, *B. anthracis* that can be treated with medical countermeasures before the onset of disease), or that can be detected by available and effective real-time or near-real-time sensor systems (such as a multitiered detection and identification system that is capable of definitive identification of an invisible cloud of *B. anthracis* released in a building). In most of the examples above, with appropriate response, exposed victims could receive medical treatment long before the onset of clinical signs when treatment is highly effective.

- **Group B—Cannot Detect, Can Treat.** An example of this group might be release of an invisible, antibiotic-sensitive, infectious bacterial agent with an incubation period of several days. For example, the *B. anthracis* release in the former Brentwood Postal Facility was unnoticed until the first victim became ill and manifested clinical signs, after which a definitive diagnosis was made. Although treatment is available for anthrax, it is of little value once the victim shows signs of illness. There are other cases in which treatment administered early in the infection or disease process could reduce the duration and severity of an infection.
- **Group C—Can Detect, Cannot Treat.** This category might be represented by an extremely fast-acting agent that causes illness or death in seconds or minutes and for which treatments are unavailable in the facility (such as cyanide) or for which there is no therapy (such as saxitoxin).
- **Group D—Cannot Detect, Cannot Treat.** The final and most difficult combination is the invisible release of an aerosolized agent for which there is no

treatment. The agent has a long latent period before the exposed victim shows clinical signs. Individuals might not know that they were exposed until hours or days after the attack (for example, the release of a well-prepared, fine particulate of dry ricin toxin). Even when the victims are diagnosed with diffuse necrotizing pneumonia and pulmonary edema, little can be done other than provide supportive care.

The four groups of generic threats support a rational approach to evaluating the options in building protection. This approach would, for example, suggest several opportunities to reduce or eliminate the impact of the powdered ricin attack scenario described above (cannot detect, cannot treat). First, a building could be passively protected from outside attack, and physical security and a personnel reliability program could be implemented to reduce the likelihood of a perpetrator gaining access to the building. Second, if the elimination of a threat agent released inside the building cannot be ensured, compartmentalization of the facility could limit the number of humans exposed to the ricin powder.

MISSIONS AND ACTIVITIES IN BUILDINGS

In planning to deploy building protection from chemical and biological airborne threats, the missions executed in a specific building need to be considered to ascertain the requirements of building protection and the importance of the mission relative to the life-cycle costs of a protection system. Addressing the full complexity of DOD missions is beyond the scope of this report, but broad types of activities can be defined to aid planning.

Broad types of activities in buildings or uses of buildings can be categorized as follows:

- *Buildings with storage only.* Assets in these buildings are limited to physical items that have utility as a resource later. Because the value of the asset is determined by later use, the need for protection varies. The primary vulnerability is from contamination of the resource that renders the storage inoperable or delays its intended use. Most threat agents do not directly damage physical objects, so the primary concern of contamination in such facilities is the safety of humans who enter the facility.
- *Buildings with equipment operation only.* Assets in these buildings are limited to operating equipment without personnel. Most threat agents (except for corrosive agents) typically would not directly interfere with equipment operations, but if the building becomes unusable by humans, the operation of the equipment could become degraded over time and affect broader operations. Risks of contamination that could affect the mission of this type of building would be the main concern.
- *Buildings with personnel only.* Assets of these buildings (for example,

barracks) are limited to people and no other activities. Because the personnel typically have other utility outside of the building, the primary vulnerability to biological and chemical threats to the missions of this building type is the secondary impact on other operations. The operations affected by an attack vary in importance, and their importance determines whether building protection should be installed to prevent loss of asset at all times (incapacitation or deaths) or to delay the loss of an asset in the short term.

- *Buildings with operations conducted by personnel.* Assets in these buildings are the sustained or intermittent activities of the personnel. The primary vulnerabilities are the interruption of operations and the hazards to personnel. Buildings with such function generally require more protection than ones with other functions because the occupants and their activities are critical to maintaining the mission of the building. The requirements of building protection are determined also by the timing of operations being conducted, which can range from intermittent to continuous.

In general, buildings are used for a combination of the above activities. Although some generalization could be made about the relative importance of the missions (for example, operations are more important than storage), counter examples often can be provided so that generalizations of relative mission importance are not useful. The importance of each building and its mission determines the goal and the level of protection required, and they all have to be considered on a case-by-case basis.

Buildings with different missions require different operational responses in combination with the appropriate building protection options (Chapter 3). For example, in a building that requires personnel to fulfill its mission, these people would require training on the operational responses in the event of an attack. Such training is not necessary in a building used exclusively for storage.

FACTORS THAT LIMIT DESIGN AND IMPLEMENTATION OF BUILDING PROTECTION

Building Procurement

Building procurement influences the strategies used in protection because implementation of protection strategies is likely to have fewer limitations if DOD owns the building. Generally, most federal and DOD space for operations can be acquired by three major avenues: build-to-acquire, build-to-lease, and lease. Building-to-acquire is accomplished by several contractual schemes, with private constructors building government-designed facilities specifically for the sole use of DOD occupants. When the build-to-lease option is used, private contractors build a facility per government design. The building will eventually be leased by the government, ideally in a long-term arrangement. Space for government use can be acquired by leasing an existing building for the sole use of DOD occupants or for shared use with private tenants.

When constructing new facilities, building protection measures can be included in the building design at the outset with a higher probability of successful installation and operation. When DOD wholly leases a facility, even if the building already exists, rehabilitation for building protection can be achieved. Although rehabilitation is a somewhat less than advantageous situation, the resulting facility can eventually function as a protected building.

If a new or existing facility is only partially leased for government operations, some lease restrictions and other limitations could pose challenges to the required protection of DOD personnel and operations in the facility. In the October 2002 report *Building Security, Security Responsibilities for Federally Owned and Leased Facilities* released by the U.S. Government Accountability Office (GAO, 2002), most of the federal government agencies surveyed responded that a major barrier to securing federal facilities arises when the space is leased. Specifically, buildings that are not occupied solely by federal employees can pose problems. Private landlords leasing space to federal agencies reportedly do not want to inconvenience private tenants sharing the space. The report gives an example of when the judiciary is assigned space by the U.S. General Services Administration (GSA) in a portion of a nonfederal office building. The security screening can be provided only at the entrance to the judiciary's assigned space without any screening at the building's entrance. In this situation, weapons and other hazardous material can be brought into the building. The report suggests that facilities leased by the government and occupied solely by federal employees could also pose problems in providing the level of security required for federal government operations and personnel. Whether building protection can be achieved in leased buildings depends on the cooperation of the landlord and private tenants. The terms of cooperation should be discussed prior to leasing and stated in leasing documents.

Building Types

In this report, building type refers to interior layout or compartmentalization. Building compartmentalization is a more useful categorization of building types than one used by DOD (DOD categorizes its buildings into "inhabited," "uninhabited," "primary gathering," and "billeting") in the context of protection from biological and chemical airborne threats. Compartmentalization within a facility generally increases building protection because it helps limit the spread of a threat agent throughout the building.

The interior space in a building can be organized into four formats. Buildings could have an open floor plan, be subdivided into cells, be designed for large assemblies, or combine all three layouts. An example of a building with an open floor plan is a warehouse. Buildings organized into cellular space contain rooms that are fully partitioned and have individual air supply or 100 percent air exhaustion (such as in hospitals and hotels). Spaces designed specifically for large assemblies of people are different from the open floor plan because assembly spaces are equipped to control the indoor environment for large group gatherings. All of these space types can be combined in one building—for example, a school building with several classrooms, open-space cafeterias, and an auditorium. Protection strategies for each of these building types likely vary with the building's mission. The ability to achieve different levels of protection depends on the building type. A highly cellular building has a reduced risk of exposing a large number of occupants to a threat agent in the event of an interior release because the threat agent can remain localized.

CONCLUSION

The need to protect a building from different threat types is driven by its mission and operations. To design an appropriate system, the goals of protection must be defined first, and the factors that limit its design and implementation (for example, building procurement and type, costs) have to be considered. Grouping of threats helps to determine the level of protection needed and the necessary components in the protection scheme to achieve that level of protection.

3

Components of Building Protection: Building Design, Technologies, and Operational Responses

To achieve the specific goals for building protection from a variety of biological and chemical threat types and to meet the requirements set by building administrators, designers, and security experts, many components can be selected. Selection of components requires an evaluation of many facility-specific details. Buildings have to be evaluated largely on a case-by-case basis because buildings vary in their "tightness": that is, their resistance to infiltration of outside air, leakiness of their air transport systems, location, degree of physical security and access to outsiders, training of the occupants, options for personal protection, and ability of surrounding resources to respond to an incident. (Figure 3-1 illustrates the complexity of planning for protection.) The relative importance of different possible outcomes of a biological or chemical attack is determined by the activities (operations or missions) in the facility. The activities also determine the required response time of the building protection to certain threats—for example, if continuity of operations in the facility is necessary, then a rapid response to fast-acting threat types is required to ensure continuous operation.

To facilitate later discussions, this chapter first discusses passive and active building protection and introduces the committee's definitions of levels of protection. Second, it reviews the options that could be used in building protection. Finally, it discusses how to integrate various protective measures to provide different levels of protection to buildings of different types and designs and considers the limitations of each level of protection.

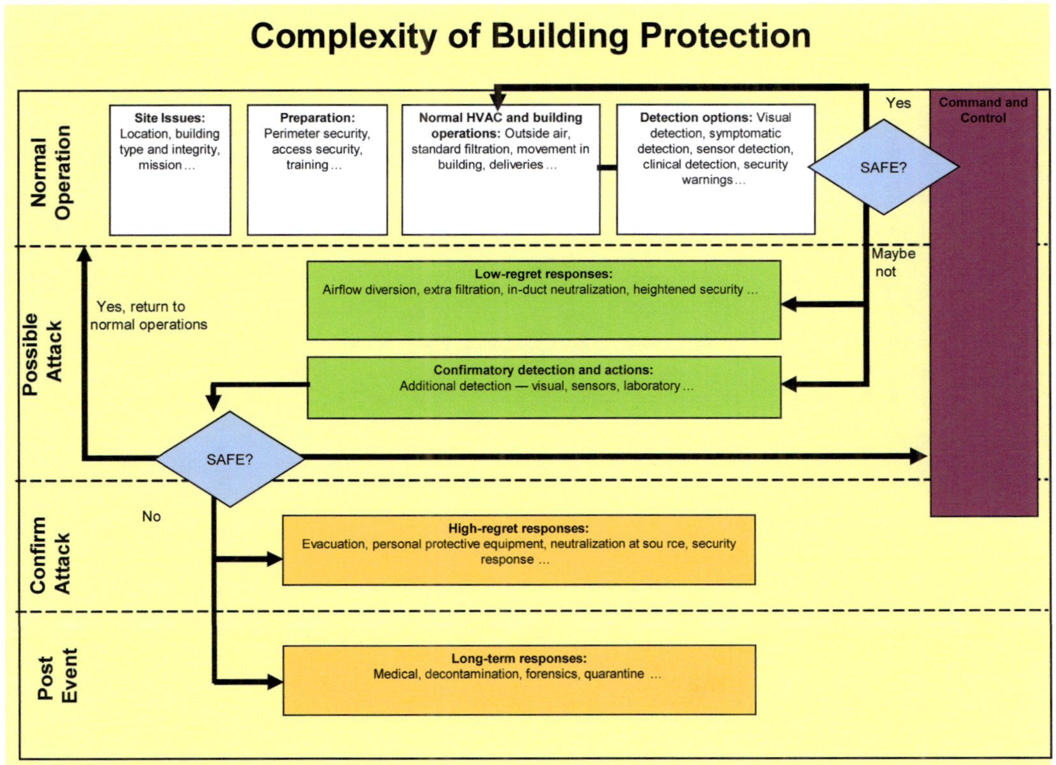

FIGURE 3-1 Illustration of the complexity of building protection (adapted from the Defense Advanced Research Projects Agency). This figure, which is not intended to be comprehensive, illustrates some of the activities and resources necessary for building protection. The flowchart does not show the different groups of threat agents and building activities that would determine the response and outcome.

BUILDING PROTECTION

Active and Passive Strategies for Protection

Incidents of terrorism that involve biological and chemical threat agents have raised awareness that buildings could be better protected from such attacks. The technologies and operational response plans currently used for building protection in most federal buildings (including military and nonmilitary structures within and outside the United States) and nonfederal buildings involve primarily "passive" approaches. Passive protection refers to systems that do not involve detectors or sensors of threat agents to influence an operational response. Passive protection utilizes the following:

- The integrity of the building as a whole to protect occupants from external threats;
- Compartmentalized spaces within the building that offer better protection from indoor releases through enhanced physical integrity of the space or from continuous local cleansing of the airstream; and
- Visual recognition of threats and their effects—directly via video monitoring or indirectly from clinical signs of the occupants—to initiate protective response actions such as evacuation, sheltering in place, mass drug administration, or donning personal protective equipment. Automated video analysis can also be used to transition visual recognition from a passive to an active system.

Although passive systems can provide protection from a variety of threat agents and scenarios, they have gaps in protection that result in vulnerabilities. These vulnerabilities fall into two categories. First, occupants could be exposed to an unidentifiable threat that goes untreated and for which therapeutic options decrease with time. An example would be an indoor release of a biological threat agent with an incubation period of days. Second, occupants could be exposed to an unidentifiable threat that is fast-acting, such as a chemical threat agent, released inside the building space. (Note that passive measures could be used to protect occupants from outside release, but only some occupants could be protected from inside release through passive compartmentalization methods.)

These vulnerabilities and their variations could be addressed by an "active approach" to protection that uses detectors to recognize the presence of a threat agent. Once detected, the threat agent could then be identified and an operational response initiated to limit the threat and allow treatment of exposed occupants. Active building protection based on sensors and detectors is not currently in wide use within the Department of Defense (DOD, 2005) because of the high initial and maintenance costs and because the risk of biological or chemical attack is low in most buildings. Test beds and current deployments such as Nord Hall of the Immune Building Program and the Pentagon will provide a basis for considering the use of sensor-based active protection (see also Chapter 5).

Levels of Protection

Nearly all buildings offer some protection from an outside release by virtue of being enclosed by walls, roofs, and openings protected by doors and windows, which limit the transport of contaminated outdoor air indoors. Buildings are typically not designed or constructed to resist infiltration of outside air entirely. In fact, some leakage or infiltration is commonly assumed to provide "fresh air" in some buildings. (For example, no ducted fresh air from an air intake louver might be provided in some simple commercial buildings or single-family residences.) Similarly, buildings are not typically built with airtight interior construction or filtration effective against the kinds of particulates and chemicals of concern in this study. All buildings are subject to widely varying quality of design and construction and to varying quality of maintenance and repair over their lifetimes. Buildings are also subject to many changes over time from aging of materials, wear and tear from ordinary use, and renovations to accommodate evolving needs and new technology. The protective performance of most buildings, therefore, is not planned, monitored, or verified. Unless the building is carefully monitored and maintained, it is unlikely to provide the protection it did when it was new, and this is a major limitation of passive protection options.

The required level of building protection from biological and chemical attacks is determined by the use of the building and the possible threats (Chapter 2). The type of building protection that can be implemented depends on many factors including the life-cycle cost of the protection system, building type, and ease of access. The committee developed the concept of four levels of protection—low-level passive, high-level passive, low-level active, and high-level active—to facilitate discussion. The level of protection is based on vulnerabilities and risks to threat agents, and a system could provide different levels of protection for different agents; a given protection system could offer active and passive protection from some biological or chemical threat agents and only passive protection from others.

Like the biosafety levels (BSL-1 to BSL-4) for microbiological and biomedical laboratories (DHHS, 2007), the four levels of protection are qualitative. The science and application of building protection from biological and chemical threat agents is not nearly as mature as biosafety in laboratories. Even for biosafety in microbiological and biomedical laboratories, the guidelines promulgated by the Centers for Disease Control and Prevention in 1984 (DHHS, 1984) are still qualitative (DHHS, 2007). Because of the variability in threat, risk, building design, and operational use of the wide array of buildings in the DOD inventory, the committee could not suggest measurable and quantitative criteria for either design or function at the time this report was written. The four levels of protection represent a plan for considering building protection. Although most buildings are designed to decrease the impact of natural disasters and fire hazards and to provide some level of indoor air quality control, they are not designed to decrease the impact of biological and chemical attacks. Therefore, some buildings have no or little protection—particularly the ones that are poorly maintained or highly porous to the outside environment. These buildings do not even reach the lowest level of protection described below.

- **Level of Protection 1 (LP-1)—Low-Level Passive Protection.** Passive protection refers to protection without the capability of actively sensing the environment for the presence of threat agents. Low-level passive building protection is based on the demonstrated protection provided by a well-constructed, well-maintained building that provides a healthy environment for occupants and operations. A building designed to provide a high-quality environment during normal operation also provides some protection from external and internal threats (Hitchcock et al., 2006). In general, an LP-1 building meets or exceeds all requirements of consensus indoor air quality standards, such as the American Society of Heating, Refrigerating and Air-Conditioning Engineers (ASHRAE) Standard 62.1, in terms of its design, construction, and operation (ASHRAE, 2004a). An LP-1 building has a well-sealed envelope to limit infiltration of outdoor air and any contaminants it might contain and to provide for minimal leakage of heating, ventilating, and air-conditioning (HVAC) air distribution to reduce unintended airflows inside the building. HVAC system types that inherently limit the spread of indoor contaminants are also LP-1 options. A sufficient supply of outside air to dilute contaminants and a moderate level of particulate filtration are also found in LP-1 buildings. Although most HVAC systems offer some degree of particle filtration, not all buildings have the high degree of filtration required for consideration as LP-1. Buildings usually do not contain adsorbents (for example, activated carbon) and absorbent filters (for example, air washers) as part of typical HVAC installations except in special circumstances. An LP-1 building has to be well maintained to ensure that protection is available when needed. The LP-1 options align with a recent study (Hitchcock et al., 2006) that recommends the above options as realistic protection of public buildings, eschewing more complex alternatives. Most DOD buildings and many public buildings have security and operational activities that are not intended for, but offer some protection from, biological and chemical airborne threats. These measures have dual-use advantages and are also part of the LP-1 options. An LP-1 building also contributes to the performance of protection systems that include sensors (that is, active protection as described below). Detectors have higher reliability and fewer false positives (a wrong indication that a threat agent is present) when operated in a clean environment provided by the LP-1 option.
- **Level of Protection 2 (LP-2)—High-Level Passive Protection.** LP-2 provides protection by further limiting exposure to intentionally released threat agents, and it is similar to MilStd Class 1 collective protection (USACE, 1999). This level of protection involves options for reducing the vulnerability to threat agents that are not part of a "standard" building system. The protective measures in LP-2 are passive because they do not actively detect the presence of threat

agents (although the operational response plans could change from passive options to active response in some situations). LP-2 options include adding gas filtration and upgrading aerosol filters and other control technologies specific to biological and chemical threats, zoning the building interior with differential pressures, relocating outside air intakes and filtering outside air, and providing local air-washes (areas that are provided with isolated, enhanced laminar airflow with local filtration at the returns). Many protection options in LP-2 serve dual purposes because they improve the working environment and the building security.

- **Level of Protection 3 (LP-3)—Low-Level Active Protection.** Active protection refers to protection with the capability of actively sensing the environment for the presence of threat agents. LP-3 offers low-level active protection and directly addresses one of the main vulnerabilities of passive systems (LP-1 and LP-2)—exposure of the building's occupants to a threat agent that is not detected and identified in time to execute therapeutic responses. LP-3 is a "detect-to-treat" option that would allow identification of a threat agent in time for treatment. LP-3 requires a broad-spectrum detection and identification system that could determine the presence of a variety of known threats within the time period necessary for an operational response. The time for detection varies by threat agent; the threat requiring the longest detection time typically involves a biological agent. Because the LP-3 option detects and identifies the threat in time only to treat the people exposed, it might not be an appropriate option for facilities that require continuous operations. Some threat agents that escape detection could have a quick impact on facility operations.
- **Level of Protection 4 (LP-4)—High-Level Active Protection.** LP-4 is a high-level active protection that addresses the second major vulnerability of the LP-1 to LP-3 approaches to building protection—the inability to mitigate an attack through timely detection. LP-4 would allow detection and identification early enough to treat the exposed victims and to make operational responses that might minimize the impact of the threat agent by preventing or limiting exposure. These operational responses might include high-regret options. (In this context, "regrets" are negative consequences of actions, as discussed in detail later in this chapter; see section titled "Operational Procedures for Protecting Buildings.") The LP-4 option is considered to be at the edge of current detection and identification technology and ability to operationally deploy. Because of current limitations in detection and identification technologies, a successful LP-4 option requires tiered levels of detection and response and uses combinations of low-regret response options with fast, nonspecific detection systems. Box 3-1 summarizes the levels of protection and the options each level comprises.

LP-2 generally has all the virtues of LP-1 and some additional passive protection. LP-1 and LP-2 are usually part of active protection so that LP-3 and LP-4 generally have all the virtues of LP-1 and LP-2. However, LP-3 and LP-4 could

BOX 3-1
Levels of Building Protection from Biological and Chemical Airborne Releases[a]

LP-1: Low-Level Passive Protection

- Select systems to minimize normal exposure.
    - Dilute indoor air and reduce recirculation.
    - Minimize leakage in HVAC system and in building (external and internal).
    - Add filtering as needed for healthy workplace.
    - Protect air intakes to reduce air contaminants.
    - Use construction methods and materials that reduce chemical exposure.
- Consider security, site selection, and operational activities that have dual-use advantages for building protection.

LP-2: High-Level Passive Protection (LP-1 + options specific to protection from biological and chemical threat agents)

- Use upgraded protection from biological and chemical threat agents compared to LP-1.
    - Upgrade filters (particulate and adsorption) specific for biological and chemical threat agents.
    - Use zoning with graded pressurization (compartmentalized).
    - Provide local air-washing vestibules.
    - Protect air intakes specifically to reduce biological and chemical threats.

be implemented without some of the basic options in LP-1 and LP-2. In general, active protection has fewer vulnerabilities, higher life-cycle costs, more complex operation, and higher risk of failure as a result of technological and operational complexity. In addition, the use of sensors also introduces the possibility of false positives and false negatives (no indication despite the presence of a threat) from the sensors. False positives are disruptive to operations and cannot be tolerated in many operational situations, so it is important to minimize false positive responses. False negatives when sensors are present could also provide an undesired sense of complacency compared to when sensors are not present (and there is also no indication of a threat).

The four levels of protection address different types of vulnerabilities. Figure 3-2 shows a cross-comparison of the levels of protection and the threat groups (selected on the basis of the ability to detect and treat the threat) they could address.

The "cannot detect, cannot treat" group (Group D) of threat agents poses the greatest challenge because of the inability to detect the threat. Including sensors

   - Include all human-in-the-loop detection options and responses, such as visual or clinical detection and the corresponding responses.
- Consider site selection, protective access control, and operational responses specific to biological and chemical threats.

LP-3: Low-Level Active Protection

- Ensure that a hidden threat agent is detected and identified in time to treat any exposed persons, essentially detect to treat.
- Provide protection for latent-acting threat agents with possible treatment.
- Include all human-in-the-loop detection options and responses in LP-2.
- Consider site selection, protective access control, and operational responses specific to biological and chemical threats.

LP-4: High-Level Active Protection

- Detect to protect (warn and mitigate).
- Include automated detection and response systems for faster reaction times.
- Use a tiered detection-response system in most cases with currently available sensor technology. Typically, low-accuracy sensors trigger low-regret responses if a threat is detected, and sensors with confirmation and identification capability are used for higher-regret responses.
- Consider site selection, protective access control, and operational responses specific to biological and chemical threats.

[a]The levels of protection could be different for different threats; for example, a system might offer LP-4 for certain chemical threat agents, but LP-3 for biological threat agents.

(LP-3 and LP-4) does not enhance protection from these threat agents because the systems cannot detect them. Hence, LP-1 and LP-2—filtration without sensors—could enhance protection from the most challenging threat types. Although detection and identification technologies will improve in breadth, specificity, and response time and expand the opportunities for the LP-4 option, passive options will continue to play an important role in building protection.

STRATEGIES AND TOOLS FOR PROTECTION

Building Design and Planning Strategies

When the built environment is to be tasked to provide protection against airborne threat agents, generalized solutions must be considered with caution because no two buildings are exactly alike, even when they have been "standardized." Unlike mass-produced appliances or automobiles, every building is custom built. Therefore, every building must be studied individually for ways to

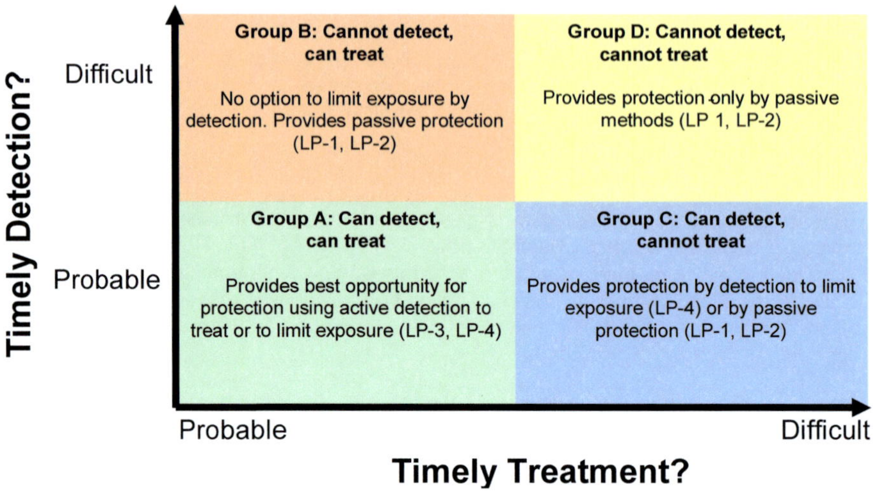

FIGURE 3-2 Illustration of how the four levels of protection (LP-1: low-level passive protection; LP-2: high-level passive protection; LP-3: low-level active protection; LP-4: high-level active protection) can be applied to the four groups of threat (A: can detect, can treat; B: cannot detect, can treat; C: can detect, cannot treat; D: cannot detect, cannot treat). In general, passive protection offers some protection from threat agents but might not be effective for all. Active protection aims to detect the threat agent in time to treat exposed victims (LP-3) or to limit exposure (LP-4).

mitigate airborne releases of threat agents. The design of a building is a product of response to functional program, climate, topography, geology, and aesthetic or iconic objectives.

In DOD facilities, the functions are classified in the Unified Facilities Criteria as billeting, primary gathering, or other DOD inhabited. In fact, these classifications might embrace everything from dwellings to offices to large armories, and more than one function could be housed within a single structure. The design responses will result in the spatial layout and selection of the structural system, HVAC, and other systems, including physical security.

Although test beds and laboratory-built spaces can be carefully controlled and ideal conditions can be achieved, field conditions for most buildings constructed for DOD and other clients vary widely from one project to another. Such variables as worker skills, false alarms due to background material, temperature and humidity conditions during construction, and materials from different manufacturers and lots can result in departures from the strict design intent and different performance characteristics for the building. If, for example, one of the mitigation strategies includes creating pressurized zones or compartments, great care must be taken to ensure that the field conditions result in airtight construction because most building assemblies are quite porous. Even different contracting methods (design-bid-build, design-build, or multiple contracts) can affect the final building results.

Buildings also change over time: they are dynamic, not static, and respond to temperature, humidity, wear and tear of use, quality of maintenance, and impacts of renovations. Modern buildings with increasingly sophisticated technologies are more than ever subjected to changes as new technologies become available. Constant rewiring or re-piping of systems can compromise the integrity, for instance, of an originally airtight partition. Therefore, it is important that designers of threat mitigation methodologies for new or existing buildings consider the field conditions that can affect construction and the likely impact of time on the original design.

From its beginnings, the built environment has had shelter and protection of its inhabitants and their possessions from natural and human-made threats among its primary objectives. It is, then, within the tradition of building design to include some element of physical security. Design strategies of an architectural nature that could be useful in mitigating the effects of airborne contaminants, including biological or chemical threat agents, are passive and belong in the LP-1 and LP-2 categories. Furthermore, the passive strategies complement the active strategies in the LP-3 and LP-4 categories.

New Building Design

New buildings can be designed for physical security—including mitigation of airborne hazards—more readily and economically than retrofits to existing buildings, especially if the security needs are anticipated early in the pre-design and design phases and are identified in a threat and risk assessment. The physical security needs, including mitigation of any biological and chemical threat agents released, become part of the functional program, budget, and design brief.

Site Selection Considerations

Design for physical security begins with the selection of the building site. A well-chosen site with access control and adequate standoffs from uncontrolled neighboring sites and rights-of-way can save costs of mitigation. However, the cost of the land could offset the savings in construction in some markets. Site selection is, of course, limited to new projects. In deciding to construct a building, the following should be taken into account to the extent feasible to achieve LP-1 and LP-2:

- Ideally, the site should be away from coastal regions subject to hurricanes and flooding, in order to minimize potential damage to the exterior envelope from winds. A damaged building envelope could affect the airtightness of the building and hence the protection from external airborne threats.

• The site should be as remote as possible from major thoroughfares such as interstate highways, main railroad lines, or navigable waterways where hazardous agents (including toxic industrial chemicals and materials) could be transported and where accidental or deliberate releases could affect the site.

• The site should be outside the landing and takeoff patterns of an airport to minimize exposure to aircraft accidents.

• The site should be located away from a neighbor or community that is potentially a target of terrorism, such as an iconic federal installation or monument.

• If the site is in a dense urban area, access to and from the site should be via streets that are not so congested that emergency responders cannot access the site quickly and evacuation cannot be effected promptly. Access from more than one street and from more than a single side of the site permits an alternative should the main access be blocked in an emergency.

• The site should be remote from and upwind of hazardous manufacturing, processing, or storage of potential airborne contaminants and far enough from combustibles so that it would not be affected by fire or explosions on the neighboring property.

Site Planning Considerations

Once a site has been selected, the designer and owner have to consider the ease with which the building can be protected from threats by the site's design. To achieve LP-2, the site perimeter can be controlled by topography such as steep slopes, berms, or ditches, or physical barriers such as walls or fences to prevent or impede access to the buildings from outside the site by unauthorized pedestrians or vehicles. Sufficient distance between buildings and uncontrolled areas outside the perimeter might be required to allow time for detection and interdiction of a threat—for example, a person approaching the building with intent to break and enter, explode a device, or introduce toxic chemicals into the building's environment. The distance from an event at the perimeter, such as an accidental or deliberate release of a toxic substance into the air or an explosion, can mitigate the effect of the event on the occupants.

At least two alternative places on the site perimeter should be provided for access and evacuation, where possible. Because entrances are inherently attractive targets, if one entrance is the site of an event, emergency response and evacuation require an alternative. For similar reasons, there should be redundancy of utility services to the site. Water, power, communications, and other utilities should serve the site from more than one point. These points of service should be protected from illicit access and potential tampering.

Another protection option is to select a site large enough to allow adequate distances between uncontrolled and unscreened vehicles and the building(s). Adequate space will be needed at the site perimeter entrances for vehicle inspection, including queuing, turnaround, and screening. On-site, keeping screened vehicles at safe distances from occupied buildings could minimize risk of impacts from an explosion or rapid intake of a discharge of toxic airborne substances, including exhaust fumes.

Fresh air intakes for buildings could be protected from unauthorized access or from accidental intake of noxious fumes, including exhaust from vehicles, by location on the building or protection with barriers.

Building Interior

The design of an HVAC system is important in building protection, but the planning and design of the building can provide many mitigating features. Limiting access points to the building to as few people as possible is one strategy to achieve LP-2. In this way, the flow of people into a building can be controlled by screening with personal inspections and sensing devices and monitored by security personnel. Access to the building could be limited and people's whereabouts in the building could be monitored. Magnetic or proximity cards and unlocking devices are now familiar in many government and institutional applications. Unauthorized people are prevented from accessing protected areas; if an employee has brought a weapon or toxin into a building and passed screening, the notion that the person has been identified and his or her whereabouts have been recorded could deter some hostile actions. At a minimum, access should be controlled, and intrusion to the following areas that could be sensitive should be monitored:

- Fan rooms
- Mechanical equipment rooms
- Electrical switchgear and emergency generator rooms
- Telephone and data panel rooms
- Filtration and pump rooms
- Interstitial spaces
- Biosafety laboratories 2, 3, and 4 and vivariums
- Laboratories and other spaces using and storing select agents as defined in Title 42, CFR, Part 73, including pathogens and toxins regulated by the Department of Health and Human Services (DHHS) or the U.S. Department of Agriculture and non-overlap select agents of DHHS
- Utility tunnels and meter rooms
- Rooftops and penthouses
- Areaways—especially if used for air intakes
- Storage for bottled gases, liquids, flammables, pressurized containers, fuel tanks, and other hazardous products
- Food storage
- Pharmacies and pharmaceuticals storage

Air intake louvers are best located out of reach or access from the ground or other uncontrolled areas. Louvers should not be located where exhausts from vehicles or other sources can be drawn into the system.

To achieve LP-3 and LP-4, certain LP-1 and LP-2 measures should be in place. For example, access to a central analysis center and to filter and air handler rooms needs to be controlled. Vulnerable areas should be isolated. Arguably, the most likely areas in buildings from which toxic substances can be introduced to the building's air are public entrance lobbies, mailrooms, and loading docks or receiving areas.

Isolation to mitigate impacts of toxic or airborne substances can be achieved physically by segregating or isolating spaces through barriers and mechanically by preventing air from such spaces being introduced to the rest of the building by recirculation. These areas are also potential sites for explosive events, so their isolation from the main building is good practice in the design of modern buildings potentially subject to hostile actions.

Threat agents can also be introduced surreptitiously anywhere in the building by someone (such as an employee) who has not been screened at the entrance. Further zoning a building to compartments served by dedicated air handlers and ductwork could minimize propagation of a threat agent in the airstream. It would also allow lockdown or physical isolation of a compartment in which a threat agent discharge is known to have occurred and allow safer evacuation of other areas. Negative pressurization of select spaces to contain contaminated air is a potential option, but care needs to be taken to avoid pressures that impair egress through doors swinging into the higher-pressure side, such as fire stairs in high-rise buildings.

Elevators, dumbwaiters, and escalators are a challenge to managing the migration of air from floor to floor. These vertical transportation devices should be placed in lobbies on each floor with automatically closing doors between the lobbies and the corridors to isolate the shafts from the rest of the floor. These are also requirements in many codes for fire and smoke control. Similarly, the ability to seal multistory atriums from the rest of the floors is essential, and codes often require them to be provided with smoke removal fans.

The porosity of modern building construction cannot be underestimated. The ability of ordinary construction to prevent air from migrating into the building, from one part of a building to another, and even into the building's interstices is not sufficient to use as a basis for passive protection against migration of contaminants. The use of plenum distribution of air, especially underfloor plenums, is particularly vulnerable to easy contamination through leakage. In the event of contamination, cleanup of plenums is much more difficult than cleaning ductwork. Plenums are best to be avoided in the design of buildings that are potential targets for toxic contamination.

In buildings with such enhanced air protection as high-efficiency particulate air (HEPA) filters, adsorption units, ultraviolet germicidal irradiation (UVGI), and other devices, rooms for air-handling equipment must be sized to allow for the generally larger-than-normal equipment and access to filters for changing. Where sampling of internal air is done using tubes to one or more central analysis centers, adequate space for running sampling tubes above ceilings and in vertical shafts must be provided with space for required bends. Service access panels must be strategically located and secured with locking hardware. Central analysis rooms must be located strategically for the areas served, and they must be adequately sized for the required sampling and analysis equipment.

Existing Buildings

Ideally, any of the measures for new buildings are suitable for existing buildings, but it could be impractical and costly to retrofit existing construction. Although site selection is not an option for an existing building, site organization and access design could improve protection of an existing building.

Relocating fresh air intakes might be possible, or existing intakes might be protected with barriers that permit the intake of air but protect against the introduction of threat agents directly into the air system. Changing air distribution and exhaust might be hindered by the unavailability of suitable plenum space above ceilings or shaft space in the building core. In addition, alternative air intakes or geographically disperse selectable intakes could be useful.

Probably the easiest protection to implement in existing buildings is access control and monitoring. Existing sites can be retrofitted for perimeter barriers and for vehicle and personnel screening at the perimeter where space and conditions allow. Other measures such as isolation of lobbies, mailrooms, and loading areas might be possible on a case-by-case basis. Wherever mitigation depends on preventing air migration from one area to another, testing and inspection of existing construction are necessary to identify inevitable leakage so that corrective action can be taken.

Heating, Ventilating, and Air-Conditioning Systems

HVAC systems are designed to provide conditioned air to the various spaces in a building in order to satisfy demands for heating, cooling, humidification, dehumidification, and contaminant removal. Consequently, such systems also represent a serious vulnerability to biological and chemical attacks in most buildings because they can function as a point of entry and distribution system for threat agents throughout a building.

The extent to which an HVAC system mitigates or exacerbates a threat agent release depends on the system's characteristics, the architectural characteristics of the building in which it is installed, and the characteristics of the release. When designing a new building or renovating an existing one, the architect and mechanical engineer have the opportunity to collaborate to reduce the vulnerability of HVAC systems to biological and chemical attacks. Alternatives for retrofitting existing systems are generally much more limited by space, operational considerations, and cost.

Recirculation Versus Once-Through Airflow

A key distinguishing feature of an HVAC system is whether or when it recirculates a portion of the air removed from conditioned spaces or exhausts all of it. Once-through systems typically provide 100 percent outside air to spaces at rates selected to meet ventilation requirements. These requirements could be as low as 10 to 20 percent of the flow rate associated with a recirculating all-air system. To satisfy heating and cooling loads at the reduced flow rate, unitary or hydronic terminals such as heat pumps, fan coils, or radiant panels might be needed.

Contaminants released indoors in one space served by a recirculating system could be redistributed rapidly to all spaces connected to the same system. The distribution of contaminants released in one space to other spaces is much more limited in a once-through system. However, pressure differences created by wind, stack effect, and HVAC fan operation can cause flows between what, in theory, are isolated spaces.

The once-through airflow system supports and complements the concept of architectural compartmentalization. However, recirculating systems can incorporate centralized air cleaners to remove contaminants collected from within a building, whereas once-through systems cannot, which makes them more vulnerable to external releases. The use of local air cleaning systems in conjunction with once-through ventilation could be necessary to provide the same degree of protection in a space once it has been contaminated. This risk needs to be weighed against overall reduction in risk as a result of more localized and slower distribution of contaminants than would occur in a recirculating system. With respect to security, the choice of a once-through ventilation system is an LP-1 measure.

Air Supply and Return Systems

In most HVAC systems, air is delivered to conditioned spaces through overhead diffusers served by a network of ducts emanating from a central air-handling unit (AHU). Unless a contaminant is distributed initially from the AHU, it can recirculate or spread only as the result of leakage to other zones. An increasingly popular alternative to ducted overhead distribution is the underfloor air distribution (UFAD) system. UFAD systems distribute air through diffusers in the floor connected to an underfloor supply plenum. The supply plenum of a UFAD system represents a greater vulnerability because the common supply could serve large areas of a building and because air from a pressurized supply plenum can leak into wall cavities and through floors to spaces below.

UFAD systems have the potential to produce better indoor air quality in buildings during normal operation (Bauman, 2003), but they could be more vulnerable to indoor biological and chemical attacks than conventional ducted air distribution systems. One of the claimed advantages of UFAD is that it supplies clean air to displace normal indoor contaminants (such as bioeffluent and emissions of office equipment) upward, out of the occupied zone. From the perspective of a biological or chemical attack, it is easy to surreptitiously introduce a dissemination device into one of the numerous floor diffusers located in uncontrolled space. The threat agent could then be distributed into a large occupied zone effectively without passing though any filters or other control devices.

Return systems could also be ducted or plenum type. Ducted returns, like ducted supplies, tend to reduce cross-contamination. Plenum returns allow return air from multiple spaces to mix, which can distribute contaminants throughout a building even following a point release and even when ducted supply is used. Return systems also could be of the ducted or plenum type. Ducted returns, like ducted supplies, tend to reduce cross-contamination. Contaminants entering a return plenum from one space could enter other spaces directly because of pressure gradients within the building or when fan-powered variable air volume boxes (which recirculate air from the plenum into occupied spaces) are used. Similarly, contaminants leaking from positively pressurized supply ductwork into a return plenum could be widely distributed. Design to eliminate supply plenums is typically an LP-1 measure.

Outside Air Supply Systems

Most large buildings are required by code to take in and distribute outside air to dilute indoor contaminants for the purpose of maintaining acceptable indoor air quality (an LP-1 option). Also, by bringing in more outside air than is exhausted by the HVAC system or other exhaust fans in the building, the interior as a whole can be positively pressurized relative to the outdoors (an LP-1 or LP-2 option). Positive pressure inside the building is beneficial in case of an outdoor release when the supply air has been properly filtered against biological or chemical agents. Relative to the supply air quantity provided by all-air recirculating HVAC systems, outside airflow is generally small. However, all-air HVAC systems frequently are configured with "economizer" controls that permit the outside air intake to be increased up to 100 percent of the supply airflow when outdoor conditions are such (relatively cool, dry air that can offset mechanical cooling) that doing so would reduce the energy consumption of the system. Some buildings also have "dynamic" or "demand-controlled" ventilation systems that adjust the outside airflow based on an indication of building occupancy, such as indoor carbon dioxide concentration.

The operational mode of the HVAC system needs to be taken into consideration when evaluating vulnerability to outdoor releases. Under some conditions, economizer controls increase outside airflow to 100 percent, which could be four or five times the minimum required for ventilation. The consequences of an outdoor release when an economizer operates the system at 100 percent outside air could be much worse than when the same system experiences the same release while in minimum outside air mode. If coordinated properly with event detection, however, the ability to bring in large quantities of outside air to dilute indoor contaminants rapidly has protective value.

Outside air controls intended to establish an operating mode that would mitigate the effects of a release are an LP-2 option if the threat agent is detected by observation and response is initiated by a human; they are an LP-3 option if the threat agent is detected by sensors only in time to treat victims and response is initiated by a human; or they are an LP-4 option when the threat agent is detected by sensors and rapid response is automated.

The location of outside air intakes has been discussed in guidance on security design of buildings (NIOSH, 2003). It has generally been recommended that intakes be located high enough that a hostile person at ground level cannot introduce a biological or chemical threat agent with reasonably available means. Although proper location of air intakes can provide a certain level of protection, it contributes little to mitigating the consequences of a large-scale release, such as the rupture of a railroad tank car containing a toxic chemical or a release from an aircraft. To protect against such possibilities, outside air supply protection in the form of tight shutoff dampers—and, possibly, alternative outside air supplies or filtered outside air supplies—might be needed. The location of outside air intakes in secure locations is an LP-2 measure. Shutoff controls are part of a high-level passive response (LP-2) if the system does not include sensor technologies and a human initiates the response; a low-level active response (LP-3) if the system includes sensor technologies but the shutoff response requires a human in the loop or identification is relatively slow; and a high-level active response (LP-4) if response is linked to sensor technologies and is rapid and automated.

Air Cleaning Systems

Particulate Filtration

Filtration can facilitate removal of airborne biological particles from air as it enters a building or is circulated within a building (ASHRAE, 2004b). Key factors in determining the effectiveness of filtration include the method for capture of the return air, leakage from the building envelope, and the efficacy of the ventilation system (Hitchcock et al., 2006). The parameters affecting filtration include the size of airborne particles; the shape, electrical charge, and mass of the particles; and their concentration. Aerodynamic particle size is the most important of these parameters when utilizing filtration for particle removal (ASHRAE, 2001). Aerodynamic particle size accounts for physical size, shape, and density. Particles with a diameter greater than 2.5 μm are classified as coarse mode, while particles with a diameter less than 2.5 μm are classified as fine mode (ASHRAE, 2004b). These two modes have different control strategies. Fine-mode particles, including bacterial cells and aggregates of virus particles, are less likely to settle out of the air by gravitational settling than coarse-mode particles. Thus, the latter remain airborne for shorter periods of time. Combinations of filtration and ventilation can be used to minimize the presence of fine-mode and coarse-mode aerosols in indoor environments.

Filtration Mechanisms. There are five main mechanisms of filtration to remove particles from the air as described by ASHRAE: straining, inertial impingement, interception, diffusion, and electrostatic effects (ASHRAE, 2004b). The same mechanisms also could be used for sample collection for further analysis. The coarsest filtration is straining as particles are removed from the air by size exclusion; particles larger than the opening are retained. Impingement occurs when a particle is either too large or dense to follow the airstream around a fiber, lands on the surface, and is retained due to attraction. Impingement could occur in flat-panel and minimal-media-area filters. If there are high air velocities, the particles might not be retained because of particle bounce, unless an attractive coating is applied to the fiber material surface. Surface coatings are critical to impingement filter performance. Bag and deep-pleated filters rely on interception, which is the attraction of van der Waals forces, and require low air velocities so as not to dislodge particles that have contacted filter fibers. Erratic and random pathways of very small particles within the airstream that result from Brownian motion bring particles close to filter media for interception. As increasing numbers of particles are retained, a concentration gradient forms, and filtration is enhanced by a combination of interception and diffusion. The efficiency of this mechanism increases with decreasing particle size and velocities. Microorganisms generally are net negatively charged and are attracted to positively charged filter media. Electrostatic charge enhances the attraction of biological particles to filter media. Filters generally become more efficient as their surfaces become loaded with particles. Particle loading, however, could reduce airflow through the filter if ventilation fans cannot overcome the additional pressure drop caused by the captured material (Hinds, 1999).

Filter Efficiency, Resistance, and Dust-Holding Capacity. Efficiency is defined as the fraction of particles removed from an airstream. The differential drop in static pressure across the filter at a given face velocity is defined as the resistance to airflow, and the dust-holding capacity is the amount of dust that an air cleaner can retain when it is operated at a specific airflow rate to a maximum resistance value. The interdependence of these relationships is evaluated in a variety of test methods for filters. Most commonly used are the ASHRAE Standard 52.1 (ASHRAE, 1992), which defines the arrestance test, dust-spot efficiency test, and dust-holding capacity test; the ASHRAE Standard 52.2 (ASHRAE, 1999), which assigns ratings on the basis of removal of particles in specific size ranges; and various tests specific to HEPA and ultralow-penetration air filters. The synthetic dust arrestance test (ASHRAE, 1992) uses an ASHRAE synthetic dust standardized to consist of various particle sizes and types that is introduced into the test airstream. It applies mainly to low-efficiency filters that primarily remove large particles. The dust-spot efficiency test is based on the opacity change of a filter medium as it collects atmospheric dust (ASHRAE, 1992). It is more indicative of performance relative to smaller particles. However, neither an arrestance nor a dust-spot efficiency rating specifies performance for specific particle sizes. The newer minimum efficiency reporting value (MERV) rating of filters (ASHRAE, 1999) ranges from 1 to 20 (Hitchcock et al., 2006). Higher MERV values generally indicate higher efficiency, especially for smaller particles (see ASHRAE, 2004b).

Disposable pleated filters with a MERV rating of 8 demonstrate a 30 to 35 percent dust-spot efficiency and more than 90 percent arrestance for particles of 3–10 μm, while bag filters with a MERV rating of 12 are rated at 70 to 75 percent dust-spot efficiency and an arrestance of more than 95 percent for particles of 1–3 μm. Bag filters with MERV ratings of 13–16 demonstrate effective filtration of 0.3 to 1 μm particles and should capture 95 percent of *Bacillus anthracis* spores. Filters rated MERV 15–16 should capture 98 percent of *Mycobacterium tuberculosis* bacilli, 68 percent or more of smallpox virus, and 71 percent or more of influenza virus. Although MERV rating is not standard for HEPA filters, these filters range from MERV 17–20 for filtration of particles less than or equal to 0.3 μm (Hitchcock et al., 2006), thereby providing increased capture of viruses. Low-MERV-rated filters are often used as pre-filters in series with MERV filters rated 13 or higher. The majority of commercial buildings use filters ranging from MERV 5 to MERV 8 (Hitchcock et al., 2006). The use of pre-filters increases filter efficiency and prolongs the useful life of the higher and more expensive MERV-rated filters.[1]

The ASHRAE atmospheric dust-spot efficiency evaluation was used for high-efficiency filters (ASHRAE, 2004b), but it has been superseded or is used to complement ASHRAE 52.2 MERV ratings. The value of the dust-spot rating is based on the coloration of filter paper by ambient dust and is limited for particles in the micrometer and smaller size range. The dust-spot method incorporates the use of unconditioned atmospheric air passed through the test material. Discoloration of the air downstream is compared to upstream (unfiltered) air. ASHRAE 50 to 70 percent dust-spot efficiency filters are reported to remove 50 to 80 percent of particles 1–3 μm in diameter (ASHRAE, 1999), and ASHRAE 60 percent dust-spot filters are reported to remove 85 percent of particles 2.5 μm in diameter (ASHRAE, 2004b). The increased filter rating of 80 to 85 percent dust-spot efficiency results in removal of 96 percent of particles that are 2.5 μm in diameter.

The fractional efficiency or penetration method utilizes the introduction of uniform-sized particles into an air cleaner and measurement by an optical particle

[1]The cited MERV levels are based on typical physical characteristics (in particular, size) of various microbial species. If manipulation to weaponize microorganisms changes their physical characteristics, removal efficiencies could be affected.

counter, photometer, or condensation nuclei counter. The method provides accurate efficiency measurements but is time consuming and is primarily a research method. The penetration method that uses dioctyl phthalate smoke (USACE, 1999) or other aerosols (IES, 1992) to test high-efficiency filters (HEPA filters) is sensitive to mass median diameter and is commonly referred to as the efficiency of the filters with 0.3-μm particles (NAFA, 2004).[2] The efficiency of HEPA filters is high (99.97 or 99.99 percent), resulting in reporting of penetration values, not efficiency.

Polydispersed aerosols are used for the efficiency-by-particle size method. Upstream and downstream measurements are made using optical particle counting devices at a variety of flow rates. The dust-holding capacity of a filter is a measurement of the synthetic dust loaded onto a filter under established procedures and a measurement of the pressure drop as the loading increases.

Proper maintenance and replacement of filters is needed when a specified pressure drop across the filters is reached and at recommended intervals. Bypass of filters occurs when gaps are present around the filter material and there are channels within the filter matrix that sharply reduce filter efficiency. The higher the MERV rating, the more attention is required to ensure proper installation and maintenance of filters (Hitchcock et al., 2006).

Gas-Phase Air Cleaning

Sorbent filters remove gas-phase air contaminants using either physical adsorption or chemical sorption. Physical adsorption results from the electrostatic interaction between a molecule of gas or vapor and a surface (NIOSH, 2003), and chemical sorption results from the reaction between a molecule of gas or vapor and a solid sorbent or reactive agents impregnated in the sorbent material. A variety of sorbents are available for different applications, and they vary in their abilities to remove different chemicals.

Sorbent filters have several limitations. First, a sorbent filter could preferentially remove some chemicals in a mixture and allow other contaminants to pass through. Second, most sorbents do not perform as well under high humidity (for example, silica gel adsorbs water in preference to hydrocarbons, so that the adsorption of hydrocarbons is essentially blocked by water vapor). Third, adsorbent impregnation could lose reactivity over time, but determining when and to what extent this occurs is difficult. Fourth, sorbent filters are bulky, heavy, and expensive compared to particulate filters. Consequently, while particulate filtration is extremely common in buildings of all types and required by codes and standards, gas-phase filtration is uncommon in most buildings.

---

[2]It should be noted that the use of dioctyl phthalate smoke to test high-efficiency filters is no longer used in field certification tests due to hazards. Poly alpha olefin is used instead. For more information, see *http://www.bnl.gov/esh/shsd/SOP/pdf/IH_SOPS/IH62300.pdf*.

Filter Placement

There are numerous potential locations for filters in an HVAC system. The most common filter placement location is in the mixed airstream within an air-handling unit. Using that location results in filtration of both outside air and return air, and it protects coils and other downstream components from fouling. Filters could also be located directly in the outside airstream, at supply to individual spaces, on the return from individual spaces, in the common return, and on exhaust air (if there is concern about the consequences of contaminated exhaust). Stand-alone filtration devices that recirculate and clean air within a single zone might also be desirable.

In general, decisions about where to place filters could be based on concern about how placement would impact normal operations, maintenance, first-cost and operation and maintenance costs, and space. In many cases, accommodating the large pressure drops required by some particle and gas-phase filters is a major consideration. In the context of a discussion of protection against chemical and biological agents, effectiveness during extreme events is a high priority.

Most HVAC systems require a minimal level of defined filtration (ASHRAE, 2004a) and MERV 6, as rated by ASHRAE Standard 52.2 (ASHRAE, 1999). Filters of this performance level primarily remove larger particulate matter that could cause fouling of HVAC components and soiling of building surfaces (an LP-1 option). To a lesser extent, such filters also remove particles in the size range associated with biological threat agents. Highly efficient filtration of biological threat agents requires filters above the performance levels generally prescribed in most buildings (an LP-2 option). Specification of more efficient filters frequently increases the pressure drop through the filter bank, which results in the consumption of more fan energy. Because it is the norm to provide particulate filtration sections in AHUs, upgrading filter efficiency as part of a security retrofit is possible if existing fans and distribution systems can provide required flows under the new system operating conditions. Most buildings under LP-1 do not employ gas- and vapor-phase absorption of any kind. Such systems are bulky, expensive, and of uncertain reliability in most applications. The continuous use of such systems is an LP-2 option.

Energetic Air Treatment Methods

Other options for controlling airborne threats include a variety of "energetic" methods that use various wavelengths of ultraviolet radiation (UVC, 200–280 nm; UVB, 280–315 nm), possibly in conjunction with other air treatment modes. UVGI (primarily UVC) can be used within AHUs to irradiate airstreams and coil and filter surfaces. When UV lamps are installed in occupied spaces, they are designed to create an irradiated cavity above the occupied zone that treats contaminants that are carried into it by air currents or to decontaminate exhaust airstreams. The effectiveness of UV as a germicide varies greatly from vegetative bacteria to spores, but it could provide the best cost-benefit when it is combined with particulate filtration (see EPA, 2006, for more information). Ultraviolet radiation is also used to energize a titanium dioxide catalyst in photocatalytic oxidation (PCO) systems. It is also combined in some cases with hydrogen peroxide or with particulate filtration equipment. Like gas phase filtration, UVGI and other energetic methods are infrequently used in private sector buildings as an LP-1 option, but the market for such equipment is growing. Enhanced air treatment of all kinds is increasing in application because of a desire to provide better indoor air quality and security from biological and chemical attacks and infectious diseases. Enhanced particulate filtration and the addition of continuous operation gas-phase filtration or adsorption or energetic biological air treatment are generally LP-2 measures. Specialized air treatment and sensing and controls for on-demand use are LP-4 options.

Instrumentation and Controls

HVAC system controls are an important part of any building protection system. Controls range from global enable or disable functions to outside air (ventilation) controls and pressurization controls interlocked with fire protection systems to isolate smoke-filled areas and protect egress paths and refuge areas. Actuation of systems could be by human intervention and judgment (LP-2 option for intentional threats) or could be automatically controlled by logic on the basis of information obtained by sensors distributed throughout a building (LP-3 or LP-4 capability). For critical functions such as smoke control or response to a biological or chemical attack, appropriately located sensors, their reliability, and the rapidity and effectiveness of response are critical. The current capabilities and limitations of sensors systems for detection and identification are discussed later in this chapter and in other reports (NRC, 2003a, 2005b).

Active protection strategies such as those considered as part of the Defense Advanced Research Project Agency's (DARPA's) Immune Building Program and in LP-3 and LP-4 depend heavily on reliable sensor networks combined with effective mitigation technologies embedded in responsive HVAC systems (W. Bryden, DARPA, presentation to the committee, September 28, 2006). It has been demonstrated that when time and budgetary constraints are eliminated, the ideals of that program can be realized to some extent with humans-in-the-loop deployment, if not fully automated. However, the life-cycle cost of such systems and the maintenance required to keep them fully operational are substantial.

HVAC Zoning

The earlier discussion of architectural design focused attention on the potential value of compartmentalization. However, the ability of HVAC airflows to distribute contaminants throughout a building, particularly when the system recirculates return air, is far greater than the ability of architectural barriers to retard them. Consequently, architectural compartmentalization should be coordinated with its counterpart HVAC zoning.

The term "zone" has a variety of meanings relative to HVAC system design. A zone could be a thermal zone, such as an area whose temperature is controlled by a single thermostat. A thermal zone could include a number of physical zones (such as rooms) or could be a subset of a physical zone (such as the core area of an open-plan office space). Similarly, a pressurization zone is an area maintained at a uniform pressure that could be a single physical zone or a group of physical zones. A zone also could be defined as the area served by a single AHU. Decisions regarding how to divide a building into thermal zones are based on the balance between comfort (individual control) and cost. Decisions regarding how many separate air-handling systems to use can be made on the basis of functional subdivisions of a building, energy use considerations, or building size. Small buildings could have a single air handler, while very large buildings could have dozens of systems serving small fractions of the total floor space.

Thermal zoning, in general, has no direct relevance to building protection. Pressurization zoning to control the direction of contaminant migration is an LP-2 measure. System-level zoning serves a protective function by limiting the extent to which any single system can contaminate a building (LP-1 and LP-2 options). At one extreme, an entire building might be served by a single all-air system with recirculation, in which case a threat agent released in one space would be distributed throughout the building in a matter of minutes. (The typical all-air system circulates a volume equal to the building volume roughly every 10 minutes, and the time for contaminated air to return to the AHU and begin reaching other spaces is much shorter.) At the other extreme, every space within a building could be served by its own independent system, in which case a localized threat agent release might have little global effect on a building or its occupants.

Places of refuge and isolated zones, such as mailrooms, and active airflow control strategies can be viewed conceptually as specialized applications of architectural and HVAC zoning principles. Although the subdivision of a building into many separate systems seems to have few negative aspects, the independent systems interact with one another. Improper operation could result in unintended consequences. For example, in a building with multiple variable air volume systems, controls might be set up to maintain the area served by one system at a positive pressure relative to that served by another under conditions of design airflow. However, because the airflow required by each system varies with its heating and cooling loads and each must be met independently, it is impossible to maintain the desired relationship under all operating conditions.

At the simplest level (for example, dividing up a building into many areas served by separate systems), zoning is an LP-1 or LP-2 design strategy. The use of fixed pressurization schemes to maintain the desired flow pattern in a building is an LP-2 strategy. The creation of places of refuge is part of an LP-2 strategy if an operational plan is in place for a human to activate air treatment for the shelter; an LP-3 strategy if a threat agent is detected by sensors to enable treatment and response initiated by humans; or an LP-4 strategy if the threat agent is rapidly detected by sensors for protection, which in turn activates air treatment in the shelter.

Importance of HVAC Installation and Maintenance

Building systems, including HVAC systems, cannot function as intended unless they are properly maintained. Maintenance includes such actions as sensor calibration, testing and replacement of air cleaners, balancing of system airflows, and many other functions. Emergency operating modes need to be tested and confirmed on a periodic basis to ensure that they will function properly when called upon. The consequences of a threat agent release and the actions needed to mitigate those consequence depend on the extent to which airflows actually occur within a building and the extent to which its HVAC system differs from the system intended by the designer. For particulate air contaminants, a primary failure mode is improper installation of filters in their filter banks. Missing filters, replacement filters of the wrong dimensions that leave gaps through which large bypass flows could occur, or damaged media have been found in HVAC systems of many facilities (Braun, 1986; Ottney, 1993). Given that many building systems suffer from the consequences of infrequent or improper maintenance, the institution of a rigorous, comprehensive, and verified maintenance program could be considered an LP-1 security measure.

Building-HVAC Interactions

The effectiveness of HVAC systems in maintaining comfortable indoor conditions and in functioning properly during extreme events is tied closely to interactions with the building itself. Interactions between the building and its HVAC systems and between the building and the environment introduce a significant element of uncertainty into the prediction of air and contaminant movement. The leakiness of the building envelope and of interior partitions and floors determines the extent to which environmental impacts such as wind and indoor-outdoor temperature differences will influence indoor airflows and background aerosol conditions. Efforts to use HVAC airflows to pressurize the interior of the building are less likely to be successful when the envelope is leaky. If the envelope is sufficiently porous that it permits inflow of air from outdoors, the envelope itself could become a point of entry for contaminants released outdoors. Contaminants entering through the envelope and bypassing filters greatly reduce the beneficial impact of enhanced filtration. The use of greater care than is usually exercised in constructing and sealing the exterior envelope of a building is an LP-1 measure. Interior sealing for the purpose of facilitating zonal pressurization or reduction of interzonal airflow is an LP-2 strategy.

Detection and Identification Technologies and Associated Protection Options

An active approach to building protection (LP-3 and LP-4) requires a detection system that includes components capable of detecting the presence of threat agents and identifying them so that treatment can be administered and appropriate operational responses effected. The detection and identification system could include nonspecific triggers that initiate sample collection, turn on more specific identification technologies, or initiate low-regret responses. In tiered protection systems, the increased sampling and possible confirmatory detection would initiate higher-regret responses.

Sampling Systems

Sampling systems are the first stage in most biological and chemical agent identification systems. The basic components of a sampling system are collecting the sample, transporting the sample, filtering the sample (for example, aerosol samples are filtered to remove unwanted larger particles and debris), and concentrating the sample. The efficiency of the sampling system and sampling time needs to be considered when designing biological and chemical threat agent identification systems.

Sampling Locations and Backgrounds

Sampling for building protection can be done outdoors or indoors. Outdoor sampling is used to detect external releases, and indoor sampling is preferred for the detection of indoor releases but can also be used to detect outdoor releases. One of the challenges of detection and identification technologies is the complication of normal or intermittent background levels that can interfere with the desired measurement, either by causing incorrect signals (false positives) or by making the detector inoperable (false negatives) (NRC, 2003a, 2005b). Although there are no background concentrations of chemical threat agents, there could be low levels of certain toxic industrial chemicals (TICs) or toxic industrial materials (TIMs) in buildings; the background level of TICs or TIMs would likely be localized to situations where they are being manufactured, shipped, or used. Some current commercial off-the-shelf chemical sensors could give false alarms when they are exposed to some of the chemicals used in buildings. Natural nonpathogenic bioaerosol backgrounds are complex and vary with time and location.

Natural outdoor bioaerosol backgrounds come from a wide range of sources such as viruses, bacteria, fungi, and plants (Merill et al., 2006). Pollens from plants vary seasonally and diurnally and can sometimes result in visible clouds with more than 10,000 particles per liter of air. Weather patterns such as rain and wind also impact outdoor bioaerosol particle concentration.

Indoor bioaerosol backgrounds also come from a wide range of sources including viruses, bacteria, algae, plants, insects, animals (for example, skin scales), and fungi. Particulates in indoor air can increase greatly with foot traffic on carpets and during cleaning activities using buffers and other high-energy appliances. Outdoor bioaerosol concentrations will strongly affect the indoor air if the building includes outdoor and indoor ventilation (for example, open windows). Indoor bioaerosol backgrounds can be reduced dramatically by filtering air and decreasing the indoor and outdoor aerosol leakage of the building. In addition, decreasing the sources of bioaerosols within a building can reduce the indoor bioaerosol background dramatically, which in turn reduces failure modes of detection systems. Because bioaerosol background can vary dramatically from building to building, it has to be considered in designing a biodetection system.

Understanding the background as measured by the biodetector(s) within the biodetection system (such as particle concentration or size, ultraviolet laser-induced fluorescence [UV-LIF], and antibody-based detection) is critical. The background measured by the biodetector includes a combination of both bioaerosol background and detector interferents (aerosols that "look" like bioaerosols to the detector). To design an effective biodetection and identification system, it is critical to characterize the background using the same mechanism as the detectors within the bioanalysis system. Test beds, field studies, and deployments can be used for characterizing biodetectors under relevant operational conditions.

Transport to Detectors

Indoor sampling locations can be in rooms, hallways, or the building ductwork. When only a few detectors are used in a facility, they are often located in the ductwork at locations that provide the broadest building coverage. If the sample is taken within the ductwork, understanding airflow in the building is important to determine the time for transport of aerosols or vapors from the room, through the ductwork, to the sampler.

Sampling tubes can also be used to collect air from rooms and transport the air to one or more centrally located detectors and collection systems, thereby minimizing the number of sensing systems. When sampling tubes are used, improper design can result in substantial losses due to sorption of vapors and impaction of aerosols within the sampling tube. Important factors to consider are the tubing material, diameter, length, and geometry. Sharp bends in the sampling tube are best avoided to minimize loss during transport. Software programs are available to estimate transport losses as a function of sampling geometry and aerosol particle size (larger particles are more difficult to transport). In addition to losses within the sampling tube, the sampling time lag as a result of transport through a sampling tube has to be considered when designing a detection system.

Particle Size Considerations

Bioaerosols (and also aerosols containing chemicals) have a broad range of sizes from 1 to 100 μm; however, sampling has traditionally focused on aerosol particles in the size range of 1 to 10 μm. Particles significantly larger than 10 μm do not efficiently deposit in the lungs (alveoli), although particles as large as

100 μm can still be inspired and trapped in the nasal passages. Particles much larger than 30 μm are difficult to transport and sample efficiently because they rapidly settle and deposit on surfaces. However, because the number of individual biological organisms contained within a bioaerosol particle can increase with the cube of the diameter of the particle, larger aerosols can contain many organisms and provide sufficient sample for more sensitive detection and identification systems (for example, nucleic acid-based detection). Consequently, considerations of the complete detection and identification system, of the uncertainty of preparation of a threat agent, and of the backgrounds might result in an optimum range of particle sizes for sampling beyond the standard 1- to 10-μm range.

Aerosol Size Selection, Concentration, and Collection

Aerosol can be separated by size using devices such as cyclones, classical impactors, virtual impactors, and filters before detection. These same components can also be used for aerosol concentration and collection as described below. Other methods involving electrostatic and ultrasonic effects are being investigated for aerosol concentration and separation, but these methods are less developed and are not available for near-term deployment.

Cyclones use vortex flow to remove unwanted large particle debris from the aerosol. Ideally, a cyclone collection curve would be a step function, with particles larger than the cutoff collected and particles smaller than the cutoff passing through the cyclone. In practice, the size distribution of the collected particles has a sigmoidal shape. Increasing the size cutoff at the same flow rate will increase the physical size required for the cyclone separator. Cyclones with wetted walls can be used for the collection and concentration of aerosols into a liquid. In this case, the particles in the larger size fraction are collected into a liquid moving on the wall of the cyclone.

Impactors accelerate the particles in a jet toward a surface (classical impactors) or toward a nozzle (virtual impactors). Both approaches can be used to remove large particles from the sample airstream and typically have steeper sigmoidal cutoff curves than cyclone separators. The remaining particles can then be collected using a filter, cyclone, or impaction onto a surface or into a liquid. Classical impactors are compact but need to be cleaned frequently. Virtual impactors reduce the cleaning problem but are more expensive to build than classical impactors. Impactors have been configured to collect particles of 0.1 to greater than 10 μm with reasonable efficiency.

Filtration is used to collect all particles and is the most commonly used method when collection is done periodically (for example, over hours or days). Analysis of aerosols collected on filters requires rinsing or other methods to remove the particles from the filter. Although filtration can be very effective in collecting particles, the collected particles are not suitable for all types of analysis. For example, vegetative cells often dry out on filters, so viable organisms cannot be cultured from the filter. In addition, it takes time and different reagents to remove particles from filters for analysis. The analysis of filters has been used for detect-to-treat (LP-3) applications that do not require rapid detection.

Detection and Identification Systems

DOD has separate operational definitions for detection and identification. Detection systems detect a change in the environment that is consistent with the presence of a harmful chemical, protein, or biological agent. These systems cannot distinguish between the harmful and harmless varieties of chemicals, proteins, and organisms. In contrast, identification technologies measure specific parameters that are unique for the target chemical, protein, or biological agent. Ideally, identification technologies distinguish between harmful and harmless varieties of agents.

Many types of technologies can be employed for detection, identification, and protection against a deliberate agent release. Some systems are large, laboratory-based instruments; others are small and portable. Different approaches have their advantages and limitations. This section discusses both existing capabilities and prospects for technologies that should become available in the near future. For building protection using LP-4, rapid detection and identification of an agent release is essential. Rapid response time is required to protect occupants from exposure and to minimize the distribution and dissemination of an agent throughout the building, thereby limiting the number of exposed individuals. Detection systems for LP-3 do not require as rapid a response time for most threat agents and can use slower, but more specific and accurate identification methods.

There are two strategies for detecting and identifying agents: periodic and continuous. In periodic strategies, a discrete sample is taken at a point in time or over an interval of time and analyzed for the presence or absence of an agent. Sampling frequency is dictated by the sensitivity of the measurement (for example, collecting an adequate amount of sample for analysis could take a long time), the time needed to protect building occupants, the time for transport from the sampling location to the detector, and the time it takes to run a sample through the detector or identifier. Many analytical instruments could take several minutes to hours to make a measurement. Clearly, if the duration of detection or identification is longer than a few minutes, it is too long, and it is thus inappropriate for detecting a release and warning building occupants (LP-4). For LP-3, such a system would provide the ability to treat individuals exposed to biological agents before clinical signs appear.

Generally, detectors are used for continuous monitoring and can be placed directly at the point(s) of interest. Physical monitors detect such parameters as particle count, pressure, temperature, and light intensity, whereas biological and chemical identification systems detect the presence of specific biological and chemical analytes.

Many limitations to detection and identification of threats exist. For biological threat agents, the background level of interferents (nonthreatening biological agents that could alarm broad spectrum detectors) is often high and varies over time. Interferents could make distinguishing between an intentional release of a threat agent and variations in normal background difficult, unless the release occurs close to a detector. Even when a threat agent is detected, rapid identification of specific biological agents is difficult. The added benefit of rapid identification of biological agents would allow administrators to more efficiently mitigate the threat and prepare treatment options for exposed people. For chemical agents, rapid detection and identification is less difficult. Sensors and sensor systems that detect and identify chemical agents currently exist; however, they differ in levels of specificity and could have problems distinguishing between the threat agent and the background measured by the sensor at the location of interest. The background signal can arise from scattered photons (in optical detectors) and inherent sensor system noise or from interferent chemicals at the location of interest.

The inability to identify a previously uncharacterized agent with unknown properties is an enormous limitation of current technology. Emerging and engineered threats could potentially be detected, but identification technologies would not be able to recognize an unknown threat agent. Consequently, release of an engineered or emerging threat that current sensor technology cannot sufficiently detect and identify could denote an unknown release. Even if a release is detected and identified, planning and treatment options might be very limited. In general, many identification technologies require prior knowledge of the specific agent and will not identify agents that are not on the list targeted for identification. Therefore, there is a need for new approaches that can identify agents based upon virulence or characteristics that do not rely on a priori knowledge of specific agents. Detection technologies might detect an engineered or emerging threat because they are less agent-specific than identification technologies, but responsive actions would be severely limited without identification.

Triggers

As a verb, "trigger" means to initiate action. In the context of building protection, a trigger is a measured event that initiates another action. The trigger of a threat agent is generally more rapid, of lower cost, and less specific than the identification technology. The trigger generally has a higher probability of false alarm and does not provide the resolution of the identification system. In this context, the trigger generally does not sound an alarm but initiates a more sensitive, and usually more costly, identification system to confirm the absence or presence of a harmful compound. A trigger can also be used to initiate a "low-regret response" within a building, such as adjusting the building ventilation to minimize the spread of potential contamination throughout the building. Triggered responses are often used in tiered detection and identification systems that progress from fast, low-regret responses based on low-specificity detection to higher-regret responses based on accurate, but generally slower, identification. A tiered approach to detection and identification systems has lower risks of false alarms than the use of low-cost, less specific identification technology. The approach also minimizes the use of costly confirmatory tests.

Video Monitoring

Video surveillance is a powerful and effective method for recognizing suspicious activities. Image processing methods are being improved to the point where they can automatically identify out-of-the-ordinary behaviors from many cameras and alert appropriate personnel so that response efforts can be focused. Video surveillance also provides a strong psychological deterrent to an attack.

Another use of video monitoring is to identify when and where a release has occurred. Release of noxious incapacitating vapors can be observed as groups of people cough or pass out. In such cases, video monitoring can act as a primary detector for the presence of an attack. Video systems also have dual uses to providing feedback on the effectiveness of response options, such as evacuations, and for remote situational awareness.

Remote Sensing or Standoff Detection

Remote sensing is the ability to have a sensor at one location that can detect and identify the presence of a particular object at another. The distance between the two locations can vary widely from a few micrometers to many kilometers. However, remote sensing typically refers to separation distances on the order of meters or farther. The key aspect is that the two elements—the sensor and the element being sensed—do not come into physical contact. Operationally, remote sensing in a protected building could be used for both biological and chemical threat agents indoors and outdoors. Remote sensing includes active methods, where a remote stimulus is introduced and the response is recorded by the sensor, and passive methods that use only an ambient stimulus.

Biological Sensing. For distances on the order of a kilometer, the ability to detect biological aerosols has limited use because current methods provide only general information about the aerosol particles. These tools are used primarily to raise some awareness of a potential hazard in the area. In general, such systems utilize light absorption, polarization, or scatter to measure cloud size, shape, particle size, and fluorescence. This information can be important on a battlefield. In some building protection operations, this type of information can be used to adjust building ventilation and controls to minimize the exposure of occupants to a large external release. In general, more specific information about the nature of the aerosols is needed for building applications. Some research programs hold promise for finer degrees of fidelity in remote biological sensing, but these developing systems are years away from being operational. UV-LIF is technically a remote system because the particles and the sensing elements do not come into physical contact. Such systems require concentration of particles and are generally employed as a triggering system for further identification.

Chemical Sensing. Remote chemical sensing requires a target chemical to have absorptive energy orbitals at wavelengths that are not impacted by naturally occurring molecules in the environment. Compounds in the air, such as water, nitrogen, and oxygen, have regions of absorption that make the atmosphere functionally opaque at specific wavelengths of energy. The most common wavelengths used for either active or passive interrogation are between 3 and 12 μm. Most chemical agents have optical spectra with absorptive regions that fall outside the opaque regions of the atmosphere and, therefore, are good candidates for identification by remote optical methodologies.

Light detection and ranging (LIDAR) is a common method for remote chemical sensing. The laser can function at multiple wavelengths in a scanning mode to determine differential absorption patterns that indicate the presence of one or more chemicals in a defined optical path. The system uses the differential absorption between closely adjacent regions to identify and map the chemical cloud. Differential absorption lasers generally use a precisely defined pair of wavelengths to perform a similar function. Chemical agents and many TICs are excellent candidates for this type of sensing.

Because biological agent types express different absorption properties, LIDAR is also a candidate for remote biological sensing. However, specificity is difficult to achieve because of the complexity of the absorption spectrum and because the important differences in biological threat agents are contained in the nucleus and are not responsive to absorption. Furthermore, the wavelengths required for identification of particular agents are in the ultraviolet region of the spectrum and pose a safety hazard. Nonetheless, LIDAR can be used as a particle counter and sizing system to remotely identify the presence of a particulate (non-chemical signature-containing) cloud. Systems that exploit cloud size, shape, and particle size distribution characteristics to indicate the potential presence of a biological element exist. LIDAR systems for biological and chemical agents are sophisticated systems operating with equally sophisticated algorithms. They need frequent maintenance and have substantial power requirements. In urban areas and some other locations, aerosol background or "air pollution" is high so that the signal-to-noise ratio can be problematic. A method related to LIDAR is open-path or closed-path Fourier transform infrared (FTIR) absorption, which could potentially be used as a remote sensing method. Infrared absorption can be used for chemical sensing even though data analysis and interpretation is challenging as summarized in Vogt (2006). The maintenance, power, and algorithm development issues for FTIR are similar to those for LIDAR.

Remote chemical sensing could be used on the exterior of a building to determine the presence or absence of a chemical cloud within kilometers of the building. This capability could allow changes in the operational parameters of the building to minimize the impact of the chemical agent on the building inhabitants, particularly such low-regret options as preventing intake of outside air or overpressurizing the building with filtered air. The warning time provided would depend on the distance from target and the wind speed. Remote sensing could also be used inside a building; however, point sensors are generally more sensitive and specific. In addition, the delay between detection and contact is greatly minimized indoors because of the limited line-of-sight distances, which makes it difficult to employ active building response measures in a timely manner.

Remote chemical sensing has some limitations. Chemicals that are heavier than air might be difficult to see remotely because of building obscuration. Identification is also difficult because of possible overlap between the spectral signatures of toxic agents and those of benign compounds.

Detection and Identification of Biological Threat Agents

Biological identification systems frequently include a triggering mechanism prior to initiation of the identification system, because the identification system is expensive and it takes time to collect enough material (toxin or organism) for the system to generate a reliable and accurate identification. The most common triggers for a biological identification system are particle counters and UV-LIF triggers. Particle counters are less expensive, but they determine only the amount and relative distribution of particle size. The UV-LIF system also identifies the presence of fluorescent material (most common is tryptophan in proteins because it has a significantly higher fluorescent signature than other biological compounds). UV-LIF systems situated in facilities that also have large quantities of paper will experience inherent interference because paper particles have fluorescent whiteners within them. Other UV-LIF systems also make measurements that determine particle size or shape to reduce false positive readings. A large portion of the biological load of building air is dead human skin, which, being biological, has a fluorescent signature. The size and shape of flakes of skin are significantly different from spores or vegetative biological organisms.

Molecular methods of analysis, such as binding assays, involve identification of nucleic acids (DNA or RNA) or proteins that are characteristic of the organism. For DNA and RNA, amplification is generally required to provide an adequate amount of material for identification. Identification can be accomplished for single analyses or for small levels of multiplexing using tube-based methods such as reverse transcription polymerase chain reaction (PCR) and TaqMan PCR. For higher levels of multiplexing, DNA microarray methods are employed so that thousands of different organisms can be tested at once. For more rapid and routine analysis, immunoassays are employed for identification.

All of the molecular methods used to detect biological agents require sample collection (see earlier section "Aerosol Size Selection, Concentration, and Collection") in order to procure a sufficient amount of sample for analysis. In most cases, biological threat agents must be lysed and their proteins or nucleic acids extracted before analysis. One exception is identification that is based on the molecular recognition of proteins or other molecules on the bioagent surface using antibodies or other molecular recognition elements. For sensitive identification, the analysis of nucleic acids requires additional time for their amplification. Molecular methods for biodetection require multiple reagents, some of which must be stored at cold temperatures to prevent degradation. Consequently, molecular methods require time and laboratory-based instrumentation. Progress is being made to miniaturize and automate most of these processes using "lab-on-a-chip" methods involving integrated microfluidic systems. Even when the systems become available, they will require a collector, will not provide continuous detection, and will take time to carry out the analysis. Consequently, they cannot be considered monitors but offer near-real-time identification.

Function-based sensing is a detection scheme using materials that exhibit a response to the agent type based on some biological function in contrast to a typical binding or physical-based method. For example, living cells or tissues can be engineered to express a fluorescent protein or exhibit bioluminescence when exposed to an agent. This response is often based on an intrinsic biological function, such as apoptosis or a signal cascade. Unlike simple binding assays, function-based assays show biological relevance because they provide information about bioavailability, binding, and effect of the agent. Function-based sensing approaches can be designed to detect either biological or chemical threat agents.

The gold standard for detecting and identifying biological agents is to use culture-based assays. In this approach, a sample is used to inoculate a culture medium enabling any living organisms to multiply and grow. Culture techniques typically require 24 to 48 hours and can take up to weeks for some viruses and bacteria, such as *Mycobacterium tuberculosis,* or if the strain is unknown (which might occur for engineered or emerging organisms). In addition, culture techniques have to be conducted in a laboratory environment, so they can be used only for post-event identification (LP-3) rather than for warning or treatment modalities that require immediate action (LP-4). Culture techniques cannot be used for all biological threat agents because of the inability to culture some agents.

Detection and Identification of Chemical Threat Agents

Each chemical agent possesses a unique chemical structure with unique chemical and physical properties that enables it to be detected and identified by different detection methods. The traditional way to detect chemical agents in a vapor state is to first pre-concentrate on an adsorbent by passing several liters of air through an adsorbent column. Thermal desorption causes all of the vapors to be released from the column. A variety of methods exist for detecting these vapors. The most common commercial off-the-shelf and government off-the-shelf systems use ion mobility spectrometry (IMS). In IMS, molecules are first ionized. The gas-phase ions then migrate through a drift tube exposed to an electric field at different rates depending on their size, shape, and mass. Another detection method is gas chromatography (GC), in which the vapors are separated on a chromatography column and identified by their retention times. A thorough discussion of the characteristics of IMS, GC-flame photometric detectors, and GC-mass selective detectors as utilized in the Army's chemical weapons storage and disposal programs can be found in the National Research Council (NRC) report titled *Monitoring at Chemical Agent Disposal Facilities* (NRC,2005c).

If more definitive speciation is required, the separated vapors are introduced into a mass spectrometry system for molecular identification. These systems all take air samples at discrete intervals so that sampling is not continuous. They also are power intensive and relatively large, so they are consequently relegated to a remote laboratory or central analysis center. Such systems can be used in an autonomous fashion but can require frequent technical intervention. In addition, because of their size and maintenance requirements, they cannot be deployed in a distributed fashion throughout a building. For new construction, however, sample ports could be introduced throughout a building in which air could be pumped continuously through inert tubing to a central laboratory, where a suite of gas chromatographs would be located.

If continuous monitoring is required, vapor sensors can be used. The most common type of vapor sensor is the metal oxide sensor (MOS). In general, a MOS responds to virtually all organic vapors and provides information that a vapor release has occurred with little or no identification capability. These types of sensors have the requisite sensitivity to detect vapors at parts per million to parts per billion levels and do not require pre-concentration. Some chemical warfare agents are resistant to oxidation so they are undetectable at low concentrations by metal oxide sensors, unless the sensors are operated at elevated temperatures.

A newer technology involves sensor arrays. For vapor identification, these types of arrays are sometimes referred to as "electronic" or "artificial" noses. Sensor arrays operate on principles loosely based on the mammalian olfactory system. Multiple different sensors in the array provide differential responses to a particular vapor, and a pattern recognition algorithm can identify the agent on the basis of the collective responses. The approach relies on the use of sensor array materials with different abilities to partition vapors of interest and a transduction mechanism to measure the amount of vapor partitioned into the sensor array material. The most common array-based sensing system is the surface acoustic wave device, which uses different sensor coatings on an array of piezoelectric crystals. When the array encounters a chemical vapor, the vapor adsorbs to the sensing layers differentially and produces a characteristic pattern that is used to identify the compound. Such sensors have limited specificity. Array-based sensor systems exist in a number of research and industrial laboratories, and systems with limited functionality have been used in an operational environment. Future improvements in array-based sensors are expected to improve their sensitivity and specificity (NRC, 2005c). Chemical identification systems could, but usually do not, rely on a triggering mechanism. Systems either operate in a continuous mode or they have a pre-concentration step followed by analysis.

Deployment Considerations

In deploying detection and identification systems for building protection applications, it is important to consider the performance and life-cycle costs of the complete system rather than individual detectors or identifiers. Performance of a complete system needs to be evaluated on the basis of sampling, processing (if needed), detection, analysis, and response. The first step in designing an effective detection and identification system is determining the types of situations and scenarios for which the system will provide protection. (See Chapter 6 for a detailed discussion.) The analysis will provide response time, detection limit, and selectivity requirements for biological and chemical agent detectors as a function of placement of the detectors. The type of detectors needed will be influenced by the backgrounds within the building, so it is also necessary to understand facility backgrounds as measured by the biological and chemical agent detectors being considered for deployment. For example, particle counters could be very effective triggers for biological agents within facilities with relatively stable, low backgrounds but would not be effective triggers in facilities with high background particle counts.

Ideally, detectors and identification systems will be placed close enough to one another to be able to detect an agent with spatial and temporal resolution high enough to prevent its dissemination over a large area or redistribution throughout the facility. For example, release of a nerve agent would provide a locally high concentration that would not require the most sensitive chemical agent detector as long as the detector was in reasonable proximity to the point of release. Deployment and distribution strategies can therefore be used for effective containment of a release. Distributed sensors also provide the ability to pinpoint the release location with higher precision and might enable dispersion modeling throughout the facility to optimize response options. However, the deployment of distributed biological or chemical agent detectors and sensors is limited by currently available technologies and their cost.

Because of the limited availability and high cost of biological and most chemical agent detectors, a tiered approach could be useful in some facilities. In this case, lower-cost triggers are used to initiate low-regret responses to contain the potential release and trigger confirmation using more expensive but typically slower detectors. Clearly, the concept of operations of the system needs to be considered when designing the detection system architecture. What action will be taken when a detector (or set of detectors) gives a warning? If a detector alarm results in no action (either automated or manual), then the detector provides no benefit and funds should not be spent for its deployment.

Relative Risk of Degradation and Activities

Just as is true with the relationship between the LP-1 options and general building maintenance, the performance of all sensing and monitoring systems depends on maintenance and adherence to operational protocols. Degradation of sensor performance often occurs as the active sensing components lose sensitivity and calibration over time. Loss either of sensitivity or calibration can result in missed detection (false negative) of an agent release. Understanding and specifying the degradation characteristics of a system prior to installation are important so that replacement of components and maintenance can be performed and built into the concept of operations (CONOPS). Replaceable components need to be accessible for easy maintenance.

The context in which detection systems will operate is as important as proper maintenance. Activities that generate high concentrations of dust or high levels of vapors can compromise sensor performance. For example, particle counters used as triggers can be activated because of construction dust. The use of high concentrations of some chemicals in buildings could cause chemical sensors to give false alarms or to become saturated and lead to a false negative if simultaneous release of an agent occurs. Chemicals that have molecular structures similar to nerve agents, such as pesticides, could lead to false positives. Such agents should not be used in the proximity of detector systems. Operations such as construction, pesticide application, and cleaning activities involving solvents must be cleared beforehand to allow adequate preparations to be made to avoid false alarms or compromised sensor performance. Physical operations also can affect sensor performance. For example, if doors are left ajar, they create unpredictable airflows or cause dilution of air that could prevent material from reaching the sensor properly.

Lack of funding to support the maintenance and proper operation of detection and identification systems would rapidly lead to a loss of performance. Assuming ongoing and reliable funding is essential when contemplating system installation because the initial capital costs have to be supported by ongoing operating cost commitments. In summary, proper resources and procedures are critical to ensure that the performance of detection and identification systems remains within the necessary range to detect agents at the requisite levels.

Technological Readiness

The NRC published a report titled *Sensor Systems for Biological Agent Attacks: Protecting Buildings and Military Bases* (NRC, 2005b). That report concluded that detection and identification technologies available at that time were insufficient for providing real-time "detect-to-warn" notification of building occupants (LP-4 option). Notional detection and identification systems were postulated that provided notification of biological organisms within a few minutes (NRC, 2005b, Box 6-1). The committee is unaware of recent technological advances that provide significant improvements to technologies since the release of that report. The committee agrees with the findings and recommendations of the earlier report, which includes a detailed analysis of specific sensor technologies (NRC, 2005b).

Risk of Obsolescence of Technologies

Technologies are changing fast. A tremendous amount of research and development is taking place, and the sensor area receives particular attention. Smaller, more sensitive, and more functional detection systems are being developed in research laboratories and commercially. It is essential that installed systems have the flexibility to incorporate innovations in detection and identification without requiring massive building renovations and changes in procedures. For sensor systems to remain flexible, standardization of their implementation is necessary so that when new technologies become available, they can readily be incorporated into a building to replace obsolete systems. For example, sensors and triggers are modular. By designing a building so that sensors can readily be replaced with improved and validated sensors as they become available—without major building disruption—the protection level can be improved periodically as technology advances. In the most optimistic scenario, highly functional miniature sensing systems with onboard processors and wireless communication capabilities will become available. Because such systems require power, access to power for sensors should be built into the building design. Even with the most advanced sensors, planning for future technology changes is required in the building design phase to anticipate such developments.

In addition to new identification technology, capabilities exist to synthesize new chemical agents and to engineer new organisms. Consequently, identification systems have to be adaptable to anticipate new agents rather than limited to detecting existing threats only. In this regard, sensor arrays that enable multiple, user-defined agents to be identified are superior to single-agent identification platforms. Multiplexed arrays that have the capacity to detect additional analytes as they are identified are particularly attractive. Having sensors or arrays that detect specific classes of chemical agents is likely to be the most practical solution. For example, knowledge of the exact identity of a chemical warfare agent is not necessary for a trigger event. However, given that all G-agents effectively have similar reactivity, a functional sensor for all of those agents would be sufficient to trigger countermeasures and a confirmatory sensing (by a mass spectrometer, for example).

OPERATIONAL PROCEDURES FOR PROTECTING BUILDINGS

The protection of existing buildings and their occupants from a biological or chemical threat requires the integration of operational procedures with specific building and detection or identification system attributes and response options.

Operational protective measures span actions prior to the event (for example, maintenance of the building; security procedures that reduce the likelihood of an event; and the vulnerability of the facility) to operational procedures in response to an event that mitigate the hazard to the building's occupants or contents (for example, changing the operation of the HVAC system; evacuation of occupants). A facility manager can prepare a CONOPS for a given threat scenario from a set of specific operational response options. The operational responses should be reviewed and practiced periodically. However, uncertainties in an actual event make the operational response a complex balance of developing situational awareness and responses to protect the building and occupants. Furthermore, each of the response options affects the operation of a facility or its occupants to some extent. Some options could lead to secondary consequences, such as the possibility of deaths. The threshold below which these undesirable consequences are acceptable often determines when "low-regret" response options (for example, shelter in place in the event of an external release) are used before "high-regret" response options (for example, donning personal protection). The CONOPS needs to balance the effectiveness and potential regrets of the options and to address the uncertainties in the threat situation (type and extent of the threat). Because the development of protective action plans is presented in detail elsewhere (DOD, 2005, Appendix E), this section focuses on CONOPS when active protection (LP-3 or LP-4) is in use. An update of detailed planning guidelines for operational responses is beyond the scope of this report.

Development of an Operational Response Plan

The development of an operational response is essential because each building has its own protection system designed on the basis of its mission, location, type, protection options, and so on. Therefore, instead of developing a general plan for all buildings, an operational response plan that incorporates risk assessment and risk management approaches needs to be developed for each building to be protected (see Chapter 6). Following are the steps of a general guideline to use in developing a building-specific operational response plan in the absence of detectors (DOD, 2005, Appendix E).

1. Conduct a building survey.
2. Write specific procedures for
   - Hazard determination (threat-vulnerability analysis),
   - Decision-making process based on conditions and events,
   - Communication of emergency instructions to building occupants,
   - Evacuation, sheltering in place, ventilation and purging, and use of protective masks, and
   - Special situations.
3. Designate and train protective action coordinators.
4. Train building occupants on response procedures.

Once a building survey is completed, the process generally requires prioritization of a list of hazards—threats and their signatures (the ability to observe a threat)—usually presented in the form of credible scenarios. The signature of a threat can be the observation of the agent itself (such as a visible chemical agent or munitions) or the clinical signs caused by the agent. A signature, or its absence, largely determines the timing and ability to respond to an event. A major response bifurcation is the determination of whether the hazard is external or internal because of the different operational response options. Response to external hazards typically involves stopping the entry of outside air into the building and shelter in place, whereas response to internal hazards usually involves purging the air within the building and ultimately evacuating. Operational response plans can be constructed and executed on the basis of likelihood of threats, their entry into the facilities, response options, and their cost and benefits including possible regrets of actions. A detailed summary of the above process is found in the report *Security Engineering: Procedures for Designing Airborne Chemical, Biological, and Radiological Protection for Buildings* (DOD, 2005).

In this study, two levels of response options are added to traditional building protection plans: detect to treat after an event (LP-3) and detect to protect during an event (LP-4). It is beyond the scope of this report to develop guidelines for the implementation of response plans for new protection technologies, but some observations can be made. For LP-3 (detect to treat), because of the inherent delays in obtaining actionable information, the immediate response plans do not change; any responsive actions will be based on the previously defined signatures above. The fundamental change is the moderate- to long-term response that determines therapeutic care for occupants, situation stabilization once the threat is known, and decontamination. The addition of LP-3 to building protection is similar to what has been developed for the BioWatch program within the Department of Homeland Security (DHS), a detect-to-treat program for cities with a 12- to 36-hour response time for a specific set of biological agents (DHS, 2006). Detailed planning and response documents were developed by the cities that are using the BioWatch technologies based on the guidance provided by DHS. Similar guidance documents likely have been developed for equivalent detect-to-treat implementations in DOD (for example, the Guardian Program of the Joint Program Executive Office for the Chemical and Biological Defense).

Because of the many possible response options for LP-4 (detect to protect), the operational plan can be complex, and the plan usually involves low-regret options initiated by higher-uncertainty detection and higher-regret actions initiated based on confirmatory detection. This type of tiered response requires commensurate operational plans, which can include triggering the capture of additional information (for example, visual confirmation of symptoms or triggering of detectors with higher operating expense) and assessment of the evolving situation. Examination of multiple scenarios is critical to ensure that the operational plan is complete for the hazard assessment. Consideration of different scenarios is likely to lead to more complex response plans. Although automation of response options, such as detection triggers shutting down HVAC systems, can reduce the complexity of human response activities and provide rapid and effective response, automated components or systems cannot be realized in most facilities in the near future. Consequently, response plans with humans in the loop are still required, and continuous updating of plans and training of personnel are necessary.

Because inappropriate response actions (for example, evacuation through a contaminated part of a building or into contaminated air outside) can increase the hazard to occupants or compromise the mission of the building, resources and occupant protection require appropriate operational procedures to be developed and practiced. Even passive protection systems (LP-1 and LP-2) are more effective with some degree of operational procedures. Furthermore, developing operational plans for LP-3 or LP-4 technologies is critical to the overall performance of the system, and guidelines for developing such plans would be useful.

Personal Protective Equipment

Personal protective equipment (PPE) includes clothing and equipment used to protect individuals in their working environment from contact with infectious or toxic chemicals or physical hazards. A basic tenet of health and safety management is that collective options such as the passive and active building design elements described above are preferred to reliance on protective equipment. This concept applies also to the protection of building occupants from a biological or chemical weapons attack. Nevertheless, where collective options are themselves inadequate to protect building occupants for all scenarios of interest, protective equipment can play an important role. This section focuses on respiratory protection only because that is the primary focus of most PPE programs for building occupants.

The appropriate PPE for occupants of a facility depends on the operations conducted in the facility and the potential hazards associated with the activity. In a laboratory, factory, or research environment where hazardous materials are routinely used, PPE typically serves only as a secondary barrier for protection. In a typically nonhazardous setting, respiratory protection may be required to provide additional protection for unforeseen events, such as a toxic industrial agent spill or a terrorist attack.

Proper PPE must be carefully chosen to mitigate the hazards presented by the risk of attack and the class of threat agent of concern. In a biocontainment laboratory setting, workers would consult agent summary statements on biosafety and worker protection (Richmond and McKinney, 1999; DHHS, 2007), agent manuals (Heyman, 2005), material safety data sheets (for hazardous or potentially hazardous chemicals), facility standard operating procedures, and people (such as facility safety personnel) knowledgeable about the associated hazards to assist in the selection of appropriate PPE. In a typically nonhazardous setting, the nature of the threat agents of concern, the potential concentrations of the agents, the level of training of PPE users, and the purchase and maintenance costs of PPE are factors to be considered when selecting the appropriate PPE. Resources are available to guide the selection of appropriate respiratory protection devices (OSHA, 1999; DHHS, 2000; NFPA, 2001; NIOSH, 2004). Irrespective of the specific devices that are selected, the implementation of an active program is essential to ensure that PPE is properly selected, inspected, and maintained periodically and that users of the devices are trained.

The most likely personal emergency equipment to be used in conjunction with other modes of building protection is "escape hoods" and other respirators. Escape hoods are typically single-use transparent "bags" that fit over the entire head and contain HEPA and activated charcoal filters. Although HEPA filters used in respirators are not certified by the National Institute of Occupational Safety and Health (NIOSH) for use in a biological environment (IOM, 1999), they have been used successfully to protect personnel for many years. However, implementing a program for use of disposable airways protection equipment in a building protection system is difficult—especially in the case of a biological attack—for the following four reasons: (1) occupants have to be warned in a timely manner to don the equipment; (2) the equipment has to be tested periodically for performance; (3) there is a small chance of injury and death from use of the equipment; and (4) the equipment cost per device could be $100 or more.

SYSTEMS INTEGRATION TO ACHIEVE PROTECTION

As Figure 3-1 illustrates, the effectiveness of building protection depends on the complex interaction of multiple factors. Selection of strategies and tools for protection requires careful consideration of the threats and vulnerabilities against which to protect, the goals and objectives of protection, the procurement and type of the building to be protected, and the desired level of protection. Although technology-based sensors and detection and identification systems define active protection, their presence does not infer that LP-3 or LP-4 is achieved. Active detection systems do not enhance protection unless they are complemented by an operational response plan and operate in an environment with high air quality to maximize performance of the sensors. LP-4 protection from some biological and chemical threats, particularly the category of "cannot-detect, cannot-treat" (Chapter 2 and Figure 5-2), is not possible at this time because of the fast-acting nature of the threat agents or the technological limitations of identification technologies. Given the complexity of building protection, a tool that assists in the selection of protection options for building specifications under different scenarios would be helpful (see Chapter 6).

LIMITATIONS AND RISKS OF EACH LEVEL OF PROTECTION

Different levels of protection can be achieved using a combination of building design, detection and identification technologies, and operational responses.

However, retrofitting LP-1 and LP-2 systems into existing buildings that are poorly constructed or have marginal HVAC systems might not be possible. For example, the extent to which filtration systems can be upgraded is limited by the characteristics of such air distribution components as fans and space constraints in mechanical equipment rooms. Buildings served by a small number of large-volume recirculating systems could be difficult to compartmentalize effectively. To implement protection in buildings, an LP-3 or LP-4 system might be considered, but as noted earlier, these options work optimally in clean (LP-1) facilities and might perform poorly in other circumstances.

LP-1, a low-level passive protection option, relies on site selection, access control, construction of the building to reduce exposure, and selection of systems to minimize normal exposure. Site selection and construction are practical when the planning and construction of a new building is considered, but they are not an option for existing building stock unless surrounding areas can be altered and construction materials can be retrofitted. Access control can be retrofitted in most buildings, with a range of options from gating and guards to sophisticated screening procedures. Routine building systems provide for dilution of indoor air and reduce recirculation of building indoor air. There are minimal external (from outdoors to indoors) and internal (between compartments within the building) leakages in the HVAC system and in the building envelope. However, as stated earlier, the envelope becomes more susceptible to leakage as buildings age, and this parameter is rarely assessed in routine maintenance practices. LP-1 also incorporates particle filtering as needed for a healthy workplace, but filtration systems require maintenance and changing of filters by qualified individuals. LP-1 does not include a monitoring option and, as such, cannot provide information to document that airborne contaminants are present for possible public health response or for forensics.

LP-2, a high-level passive protection option, builds on the features of LP-1 and provides an upgraded protection from biological and chemical attacks. Upgrading of filters (particulate and adsorption) can be efficacious, but it requires monetary resources and technical personnel to routinely change the filter matrix and ensure that there is no filter bypass following filter change-out. The use of current sorption-based chemical filters often requires significant upgrades to the HVAC system because of the large pressure drop across the filter bed. The size and weight of these systems is also an important consideration. In addition, the methods to ensure adequate filter bed capacity (lifetime) are difficult, especially when chemical agents are considered. Zoning with graded pressurization, local air-washing vestibules, and protected air intakes are necessary retrofits for existing buildings and important considerations in the design of new buildings. None of the features of LP-2 buildings provides a monitoring capability.

LP-3 is a low-level active protection option that ensures that a hidden threat agent is detected and identified. LP-3 is a detect-to-treat option. Although LP-3 provides protection for latent-acting threat agents with possible treatment, it might not capture information quickly enough for fast-acting threat agents. LP-3 can be a fully automated detect-to-treat system and can also include human-in-the-loop decisions and actions to activate systems—that is, the sensor system notifies a person of a potential biological or chemical attack and the person takes action to minimize impact of the release. Thus, LP-3 might require training and sophisticated operational procedures to minimize human error. Another limitation of LP-3 and LP-4 is that active detection has the potential of falsely indicating the presence or absence of a threat.

LP-4 is a high-level active protection that can "detect to treat," "detect to mitigate," or "detect to warn and protect." LP-4 includes rapid, automated systems. LP-4 eliminates the human decision factor, but the complex and sophisticated automated systems require routine maintenance to ensure their proper operation.

The relative risks of different levels of protection in the event of a biological or chemical attack depend on multiple factors. The fundamental risk underlying all levels of protection is the risk of exposing occupants to harm and disrupting building operations. One factor that influences this fundamental risk is the environment in which the protection system operates. If an active detection system (LP-3 or LP-4) operates in an environment with high background that causes either high false positives or negatives, an LP-3 or LP-4 system might not offer additional protection compared to an LP-2 system. The relative risks of different levels of protection can be assessed only in the context of the building in which the protection system operates.

4

Metrics and System Evaluation

The ability to protect occupants, activities, contents, and buildings themselves is the overarching goal of the different architectures for building protection. Several measures of effectiveness have been developed and used to assess protective ability. The appropriate choice of metrics and criteria for evaluation depends on the goals of protection and the objectives of the system design. The goals could be protecting resources (personnel, contents, or buildings) and meeting mission requirements (ensuring continuity of operation). The objectives of a system design could be low maintenance and service of system components or maintaining a preset budget. Preset metrics and criteria for evaluation are key elements in assessing whether a building protection system achieves its protection and operation goals and meets its design objectives.

METRICS

Metrics are measures used to assess and compare the performance outcomes of different systems. The metrics of building protection systems are measured indicators of the impact of a protection system on the occupants and operations of a building in the event of an attack, and most common ones include protection and operational performance metrics.

Protection Metrics

In building protection, the primary focus is on protecting building occupants and contents. Protection metrics provide a quantifiable measure of a system's efficacy in protecting building occupants and contents. The degree of protection offered by a system can be assessed by comparing protection metrics in a building with or without (or before and after deployment of) a protection system. According to some prior and existing building protection studies, the protection of people could outweigh the protection of contents. Some commonly used protection metrics seen in prior and existing building protection programs include fraction of building exposed (FBE), fraction of occupants exposed (FOE), and lives saved.

The Immune Building Program of the Defense Advanced Research Projects Agency (DARPA) used FBE as a primary metric. FBE is defined as the fraction of the building (by volume) in which occupant exposures would exceed a prescribed level or guideline, typically evaluated as a function of the mass of agent released.

For chemical warfare agents and most toxic industrial chemicals, the exposure criterion is the acute exposure guideline level (AEGL) (NRC, 2000, 2002, 2003b, 2004, 2006) or other similar estimate of acute toxic potency. For biological agents, the exposure criterion is the infectious dose, that is, the number of organisms believed to be necessary to overwhelm host defense mechanisms and establish an infection.[1] FOE measures the fraction of occupants that are exposed to a threat agent. For this metric to be useful, the amount of exposure at a given time or duration and the type of exposure (for example, skin, inhalation, or ingestion) need to be specified. FOE then can be derived in a similar manner, provided information is given about the location of the occupants and their exposure levels. FOE is not commonly used as a primary quantitative metric, but it is inherent in some of the chemical and biological protection architectures that have been developed and deployed. (See Chapter 5 for specific demonstrations that use FOE and FBE metrics.)

FBE relies on an experimental measurement or analysis of the transport and dispersal of agents or toxic materials within the interior spaces as a result of releases either outdoors or within the building. For example, DARPA's Immune Building Program uses multizone contaminant transport modeling as a principal means of estimating the concentration and exposure profiles within the interior spaces. Several such models are available; the Immune Building Program adopted the use of the CONTAM[2] multizone code (NIST, 2006a; Walton and Dols, 2006). Tracer experiments conducted as part of the Immune Building test bed program have shown that the modeling approach can provide results comparable to actual measurements. As is the case with most models, accuracy is highly dependent on an adequate understanding of the input parameters and inherent model limitations.

FBE and FOE have several limitations. They do not take into account the

[1]Infectious dose might not be the optimal exposure criterion to use because it is highly dependent upon the methods used for agent preparation and the strain of agent used.

[2]CONTAM is a multizone airflow and contaminant transport analysis software.

time between initial release and when the exposure guideline is exceeded (if this occurs). A building can be evacuated before the threat agent released has spread in the building. Another limitation of either FBE or FOE is that not all spaces within a building require equal protection. This is especially true for buildings that are operations centers where the highest value is protection of continuity of operations. FBE and FOE also assume that the occupants or the state of connectivity in a building is static even though personnel move around the building and the opening and closing of doors could change the connectivity of a building. They also do not take into account whether and when the diseases caused by exposure can be treated.

Another protection metric, which is related to FOE, is the number of lives saved or the reduction in the number of exposed occupants as a result of protective architecture. If the disease progression of a threat agent and the countermeasures are known, then the number of lives saved or reduction of exposed occupants as a result of building protection can be estimated. Number of lives saved or the reduction of exposed occupants can be measured as decreased mortality, morbidity, or costs associated with human health effects.

Operational Performance Metrics

The operational aspects of building protection are important considerations, even though they are less quantifiable than the protection metrics. The need for continuous operation has an important influence on protection system design. If part of the protective response is building evacuation or movement of personnel to an interior shelter, essential operations could be disrupted. Key activities that cannot be disrupted must be accounted for by the protective architecture. Similarly, the tolerance for false alarms could vary from building to building depending on the need for continuous operation.

If part of the protective architecture requires an active response from occupants, such as evacuation or seeking shelter, then the response time can be an important criterion. The paucity of data on response times does not allow realistic estimates to be made. The overall performance metric analyses need to ensure that response times are estimated cautiously and with full awareness of the impact of their variability on the overall performance.

The time for recovery, restoration, and return to service is linked to the issue of operational continuity or future use of a material asset. If there are alternative means to meet operational criteria (for example, personnel at different sites or buildings can maintain critical operations off-site), then return to service might not be a major consideration. Nonetheless, a protective architecture that results in the temporary loss of an operational asset for a few days could be desirable when restoration to service of a building without protection takes months or years.

Operationally, user acceptance can affect the performance of protective systems, especially those in which some response action by personnel is necessary (for example, evacuation and sheltering). Acceptance by users also affects maintenance of the building protection system because users who see the value of a system are more likely to maintain or replace its components or comply with rules and regulations set to ensure its efficacy. If a protection system is to be implemented, the benefits of the system and its proper operation need to be communicated to users and even to the building occupants.

There could be ancillary benefits for some building protection systems, such as improved air filtration leading to higher-quality indoor air. Recent research suggests that air quality is correlated with sick days, health care costs, and productivity, but it could take some time before such measures are sufficiently reliable to be useful in a cost-benefit analysis. In the near term, however, the likelihood that such trends exist and are statistically significant might serve as an ancillary motivation for security improvements (Fisk, 2000a,b; Seppänen and Fisk, 2006).

MAINTENANCE AND COST EVALUATION

In addition to quantifiable protection metrics, service, maintenance, and cost issues must be considered in protection system evaluation. Compliance with cost and maintenance criteria can be used for decision making and to assess whether the maintenance and budget objectives are met. During the design phase of a system, preconceived standards for the service and maintenance of its components and a budget for the total cost of the system are set. The evaluation phase assesses whether these standards and the budget are met; if not, the system or the design plan could be adjusted. Even if adjustments cannot be made, the evaluation provides valuable lessons learned to guide future protection efforts.

Service and Maintenance

As with operational metrics, maintenance issues have important consequences for decision making. In many cases, maintenance has an impact on costs, particularly over the lifetime of the building. Almost all mechanical and electronic systems require periodic maintenance to ensure that they are performing as originally expected or required. These maintenance activities, such as periodic calibration and performance testing, incur costs including labor, supplies, and consumables. Because maintenance of protection systems is often essential to ensure that they work properly, additional training and staffing beyond normal building or heating, ventilating, and air-conditioning (HVAC) system maintenance are needed. For example, installation of improved filtration requires greater care to ensure that there is no air bypass around the filters. In some situations, follow-up aerosol testing is needed to verify system efficacy.

In addition to routine maintenance, lifetimes or failure rates of the protection system components influence decisions about the types of systems to deploy.

Some of the criteria that have been used for evaluation are mean time to failure or system or component durability. A related issue is the ease and cost of component replacement. System or component obsolescence should also be taken into consideration. Obsolescence is related to the adaptability of the overall system to changes in some of the components or changes in protection requirements.

Cost Evaluation

It is probably safe to say that the more people or interior area a building protection system saves from exposure to a release, the better the system is. It is not plausible to say unequivocally that the cheaper the system, the more desirable it is, or vice versa. For cost to be used as a criterion for evaluation, it has to be considered in terms of benefits received. This is known as cost-benefit analysis and is used frequently in the decision-making process by federal agencies (White House, 1992). A detailed discussion of cost and its role in risk assessment, risk management, and decision making can be found in Chapter 6.

The methodology for performing cost-benefit analyses is complex and beyond the scope of this report to discuss in detail. However, general principles can be adapted from *Cost-Benefit Analysis Guide for NIH IT Projects* (NIH, 1999), even though it is directed at selecting information technology systems, and from the *Standard Practice for Measuring Life-Cycle Costs of Building and Building Systems* (ASTM, 1999) and *Measuring Benefit-to-Cost and Savings-to-Investment Ratios for Buildings and Building Systems* (ASTM, 1998),both published by the American Society for Testing and Materials. Cost-benefit analyses, as generally performed by federal agencies in compliance with Office of Management and Budget Circular A-94 (White House, 1992), typically include comprehensive estimates of the projected benefits (such as lives saved or reduction in FBE as discussed earlier in this chapter) and costs for all alternatives. Cost-benefit analyses are performed before and during design of a system to optimize design performance and expenditure (Chapter 6). However, they also could be performed after implementation to determine whether the expected benefits are achieved within the preset budget. To assess whether budget objectives are achieved for building protection designs, costs have to be ascertained over the system's life cycle.

Components of cost for protection include the initial cost, cost of operation, and cost of maintenance. Initial cost includes design fees and cost of personnel (such as building operators) and subject matter experts (such as architects, engineers, and risk assessment and management experts) who dedicate their time and labor to plan the protection system, cost of required building modifications, and cost of equipment, installation, and commissioning. Operating costs include all impacts on building operation staff and their necessary training and on utility use such as incremental HVAC fan energy use associated with higher-pressure-drop filters or the cost of operating lamps in an ultraviolet germicidal irradiation system. Maintenance costs include replacement of used parts such as fully loaded filters, periodic calibration of instrumentation, and testing to confirm proper system function. Collectively, these cost factors determine the lifetime cost of ownership.

A life-cycle cost analysis frequently is used to justify a larger initial cost to obtain the benefits of lower operating or maintenance cost. In the case of security enhancements, consideration of the system life cycle takes on increased importance because failure to commit to the ongoing costs of maintaining such systems will compromise their ability to perform as intended.

Representative cost information for particulate and gas-phase filtration systems has been published by the National Institute for Occupational Safety and Health (NIOSH, 2003). This source cites a range of \$6 to \$40 per square foot from continuous high-efficiency particulate air (HEPA) filtration and activated carbon filtration to sensor-activated military filtration systems. Associated operating costs (primarily the cost of moving air against the higher resistance of such filters) are estimated at up to \$2 per square foot each year, which is comparable to the energy cost incurred by a typical commercial building (NIOSH, 2003). HEPA filters are roughly 10 times as expensive as standard efficiency particulate filters of the same size, but the cost of gas-phase media can be an order of magnitude more expensive than a HEPA filter. Site factors such as space limitations and fan characteristics can add cost or limit the range of options, particularly in the case of retrofits. Qualitatively, these data indicate that building protection can be costly and that cost-benefit analyses are, therefore, important in justifying the costs of protective measures. Some passive security measures, particularly if implemented in a new building design, might carry no first-cost penalty and could reduce operation and maintenance costs. An example is architectural compartmentalization combined with the use of a dedicated outside air (once-through) system for ventilation.

CONCLUSION

The appropriate measures of effectiveness and criteria for evaluating a building protection system depend on the goals and objectives of building protection. Although the focus often is on the number of lives saved, considerations related to operational performance, maintenance, and cost should not be overlooked when planning for building protection and should be used to evaluate system-wide risks and benefits.

5

Analysis of Current and Prior Building Protection Programs and Studies

Although protection of buildings and building occupants from biological and chemical airborne threats has received increased attention over the past decade, few projects to date have incorporated a set of integrated building protection strategies. All the systems considered in this report have passive protection elements, but some of them complement passive protection with active response or control approaches. These few projects provide important insights into future development of building protection approaches and architectures. Each system, however, is uniquely tailored to address specific protection goals and, thus, has different levels of performance. Therefore, broad conclusions cannot be drawn about the economic feasibility or general applicability of the systems to other buildings.

DEMONSTRATION AND IMPLEMENTATION

Smart Building (Demonstration)

Description

In 1999, the Defense Threat Reduction Agency (DTRA) commissioned the design, development, and implementation of the Smart Building program, with the goal of demonstrating a comprehensive biological, chemical, and radiological protection system. The demonstration focused on the building that housed the Olympic Coordination Center (OCC) of the Utah Olympic Public Safety Command (UOPSC) and the Federal Bureau of Investigation's Joint Operation Center (JOC). UOPSC was a multijurisdictional entity that consisted of local, state, and federal organizations responsible for emergency response and law enforcement during the 2002 Winter Olympics in Salt Lake City, Utah.

Building protection features were retrofitted (the building was about four years old at the time) into a six-story commercial building totaling approximately 16,700 m^2 (180,000 ft^2) of floor space plus three levels of underground parking. OCC and JOC occupied the fifth and sixth floors of about 5600 m^2 (60,000 ft^2). A detailed description of the Smart Building is given in a six-volume series of reports (see Allen et al., 2006, for an executive summary and listing of the reports). The protection features of the Smart Building system were in place from two months prior to the 2002 Winter Olympics and the Paralympics to two months after the events. The system protected about 50 occupants on the fifth and sixth floors, and its estimated total cost was $22.2 million over its four years of operation. The protection system has since been dismantled and the building restored to its original configuration.

The building protection system consisted of collective protection (CP)[1] for the fifth and sixth floors and a multiple sensor system that could trigger a change in ventilation system for the first to fourth floors. Therefore, the Smart Building protection system for the fifth and sixth floors can be regarded as a high-level passive protection system (LP-2). Although the heating, ventilating, and air-conditioning (HVAC) system for the first to fourth floors included only standard particle filters and no gas-phase chemical filtration, the sensor-activated responses were LP-4. In addition, physical security systems were installed and response training and plans implemented for the entire building. Entry to the fifth and sixth floors involved additional security and passage through a decontamination air lock in the event that contaminated (or potentially contaminated) personnel had to gain entry to OCC.

The key elements in the implementation of the CP system for the fifth and sixth floors were building modifications to eliminate air leakage into those floors and the mechanical system of the collective protection air handlers that supply chemical- and aerosol-filtered air. Building modifications that provide LP-1 and LP-2 focused on finding and, to the extent feasible, eliminating leakage between the CP floors and the outside and between the CP region and the floors below. These modifications reduced the airflow requirements to maintain the positive indoor-outdoor pressure gradient and the infiltration from the floors below. The existing HVAC system was also modified so that it serviced only the lower four floors. An entirely separate air-handling system for maintaining the CP overpressurization was installed to provide about 560 $m^3\ min^{-1}$ (20,000 $ft^3\ min^{-1}$) of air that was filtered by a high-efficiency particulate (HEPA) air filter and activated carbon filter units to maintain the CP overpressurization units. The design requirement was to maintain overpressurization at the building shell for wind loads

[1]Collective protection is the provision of a contaminant-free area where personnel can function without individual protective equipment, such as a mask and protective garments (DOD, 2005).

up to 6.7 m s^{-1} (15 mi h^{-1}). The CP HVAC system was entirely once-through (see Chapter 3), with heating and cooling loads handled by separate space-conditioning systems on each of the two floors. These were air-based systems located in a room on each floor where the filtered outside air was mixed with recirculated air from the rest of the floor (no deliberate cross-floor air circulation). All the space conditioning was handled by these units. Thus, all the conditioned air delivered to the fifth and sixth floors was exhausted through bathroom exhaust systems or through incidental leakage paths to the outdoors or to the floors below.

Protection of the first four floors, which were occupied by other government and nongovernmental tenants, relied on physical security at the ground floor entrances and HVAC control procedures that could be initiated by an alarm from one of the chemical or radiation sensors deployed within the building and at external sites in the vicinity of the building. The chemical and radiation sensors were commercial off-the-shelf systems. The typical building ventilation response options—programmed for automatic initiation by the sensors—were an HVAC purge mode if an internal release was identified and HVAC system shutdown for an external detection. There was also a similar, operator-initiated response capability that could be implemented on the basis of information obtained from law enforcement and other agencies.

The biosensor and response system used a tiered approach. Samples of airborne biological material were taken continuously by the Joint Biological Point Detection System (JBPDS) equipment, which was not commercially available. Operation of these systems was monitored continuously at a separate control facility. Initial detection by the JBPDS equipment was followed by a second-tier assay analysis that, if positive, would result in physical collection of sampled material for additional analysis off-site. The confirmed second-tier analysis would result in notification of JBPDS leadership and shipment of a sample to the Utah Department of Health for further laboratory analysis. Because the confirmatory test would take 12-24 hours to complete, the primary function of the higher-tiered detection was to define treatment and decontamination responses (LP-3 options). Change in building or HVAC operation on the fifth and sixth floors would not be initiated until the Utah Department of Health analysis confirmed detection of a biological threat agent.

Discussion

The principal requirement of the Smart Building program was to ensure continuous operation of essential functions by OCC and JOC. Building protection for operational continuity was achieved primarily through LP-2-type options—continuous overpressurization with filtered air of the fifth and sixth floors. Treatment and decontamination responses were provided by LP-3-type options. Physical security (LP-1- and LP-2- type options) at the building and at the entrances to the fifth and sixth floors provided additional protection primarily to minimize the threat of a biological, chemical, or radiological attack within the CP zone. Assurance of continuity of operation of OCC was critical. A standby emergency generator was included to provide power for the CP HVAC system and for continued operation of the OCC functions. If key personnel were outside OCC at the time of an attack, airlocks and a decontamination procedure (LP-2-type options) were available to provide safe entry.

Because OCC was in operation for a short duration and because of its attendant protection requirements, few data on the long-term operating demands and costs of any of the systems were collected. The summary report (Allen et al., 2006) noted that there were challenges to retrofitting a protection system into an existing building. For example, integration of the Smart Building control system with the existing building management system was not seamless partly because some elements of the existing control system were designed to prevent damage to the HVAC system.

Washington Metropolitan Area Transit Authority

Description

The Washington Metropolitan Area Transit Authority (WMATA) has installed an early warning crisis management system to detect releases of threat agents in the subway (metro) system to aid immediate and medium-term response. Initiated in 1997, the protection system has been a cooperative effort among WMATA and the U.S. Departments of Energy, Transportation, Justice, and, more recently, Homeland Security (DHS). Much of the technical work has been conducted by Argonne and Sandia National Laboratories. The collaboration established the Program for Response Options and Technological Enhancements (PROTECT). A summary of PROTECT is given in Campbell et al. (2004).

The PROTECT architecture consists of sensors deployed in various subway stations, complemented by closed-circuit television (CCTV) cameras that have automated and manual pan-tilt-zoom capabilities. These sensor and camera combinations provide data continuously to a centralized chemical-biological emergency management information system (CB-EMIS developed by Argonne National Laboratory) located in a centralized WMATA operations control center. In addition to the sensor and video data from the stations, train operation data and ambient meteorological data are also ported to the CB-EMIS system. Under normal operations, CB-EMIS can provide operator access to the multiple fixed and movable cameras throughout the metro system to assist law enforcement officers or firefighters. It also monitors the status of the sensor systems deployed in the metro.

When a sensor alerts, the signal is provided both to the CB-EMIS system (where a visual indicator is displayed in the operations control center) and to the movable camera system(s) in the station, which focus on predetermined locations associated with sensor locations. The video observations are then used by personnel in the operational control center to determine whether station patrons are in distress, thus providing confirmation of the sensor alarm. The linkage between the sensor alarm and the video camera motion is the only automated element within the response system.

In addition to providing situational awareness, CB-EMIS contains a belowground and aboveground dispersion modeling capability. Information on agent concentration obtained from one or more sensors is combined with information on train movement and the ambient meteorology to estimate the location, strength, and fate of the release. This information is also used to estimate the transport and dispersion of the threat agent within the subway system and the aboveground dispersion as a result of emissions from the station entrances and vent shafts. The model outputs are used to identify the hazardous zones within and adjacent to the subway system and are updated as additional data and information become available. To provide situational awareness to first responders on-site, communications access stations are located outside the subway stations.

Establishment of predetermined response strategies to support decision making by WMATA and other emergency responders during an incident is key to the operation of the PROTECT system. Response options include stopping trains (some or all) or moving trains away from the affected areas. The video information also can help in the deployment of emergency response personnel to the affected station(s).

Discussion

PROTECT, as currently implemented for WMATA, relies on "human-in-the-loop" response and decision making, except for the initial automated camera response triggered by sensors. Verification of detection, performed by staff in the operational control center (alerted to do so by the sensor alarm), is a necessary first step before any responsive actions are taken. CB-EMIS has been developed as a situational awareness tool where all event and supplemental (such as data on hazardous chemicals) information can be accessed and displayed. CB-EMIS also provides estimates of threat agent dispersion, the location of hazardous areas, and predetermined response strategies. These functions provide important inputs to the response decision making.

Immune Building Program Demonstration

Description

In 2001, the Defense Advanced Research Projects Agency (DARPA) initiated the Immune Building Program with the goal of making military buildings and their occupants less attractive targets for attack by biological and chemical threat agents. DARPA recognized that one of the most difficult aspects of the effort was protection from agent releases within a building because of the small amount of mass required for a successful attack (compared with most external attacks) and the possibility of direct exposure of occupants before the threat agent could be removed by mitigation approaches such as filtration. In addition to protecting occupants, the objectives of the Immune Building Program were timely restoration of service and the preservation of forensic evidence. In establishing this program, DARPA recognized that active response protection of building occupants (LP-4) had not been demonstrated in the context of biological and chemical threats to buildings. Full-scale end-to-end tests, data, and models to examine various approaches and trade-offs did not exist. Moreover, many of the required technologies or components were unreliable or not yet available at the time the program began.

The Immune Building Program was started with two parallel efforts. Phase 1 included a set of analysis and modeling studies designed to define the problems, issues, and their scope. It also included a development and demonstration program for new technologies that could be tested later in Phase 2 test beds or deployed as part of the operational demonstration. Concurrently, a modeling and simulation tool resource called the Building Protection Toolkit (BPTK) was developed to reduce risk in the design phase and to optimize strategies, components, and concept of operations (CONOPS) in the test bed.

In Phase 2, full-scale experiments were conducted at existing (but extensively modified) buildings at the Nevada Test Site and at the decommissioned Fort McClellan U.S. Army base near Anniston, Alabama. The Phase 2 tests were designed, in part, to examine combinations of passive and active control strategies to prevent or reduce occupant exposures from an internal release. The goal was to have an optimized design that could serve as the basis for the operational system deployed as part of the operational demonstration. Phase 2 tests provided an opportunity to collect experimental data on the dispersion of particles and gases in indoor spaces and to compare the data with indoor dispersion models. Phase 2 also provided a limited evaluation of new technology developments.

In Phase 3, the final element of the Immune Building Program was an operational demonstration of the system in an occupied military building under real-world operation conditions. The selected site was Nord Hall, a building that houses certain functions of the U.S. Army Chemical School at Fort Leonard Wood, Missouri. At the time this report was written, full deployment and efficacy testing had not been completed, nor had CONOPS been fully developed and tested.

As was the case for the DTRA Smart Building program, deployment of the Immune Building system was retrofitted to an existing building (built in the mid-1990s). The main components of the protection system at Nord Hall are upgraded absorption and particle filters on the HVAC system (LP-2 options) and active HVAC pressure and airflow control triggered by the sensor system to provide protection from internal releases. Because of the limited availability of interior space and the costs of modifying interior spaces, one of the outside air-conditioning units for incoming air was located at ground level outside the building. This location is not desirable from a security perspective. The air treatment systems included standard military off-the-shelf technology for particle and gas filtration. The passive system provides protection against external releases. However, the building itself is not regarded as being collectively protected (that is, it does not maintain a positive pressure gradient—inside to outside—everywhere across the building shell). The active system is linked to a series of biological and chemical sensors that are set to take air samples continuously from the various zones within the building. The sampler and response architectures are described below.

A centralized equipment room houses all of the biological and chemical sensor systems. Tubing was installed throughout the building to deliver air samples from various locations continuously to the equipment room. Studies were done to determine the acceptability of the air transit time relative to the other time delays in the sensor system, such as sample processing. Transport efficiencies for both gas and particle samples between the sampling sites and the sensors were also assessed empirically. Overall, these studies considered the trade-offs of cost, transport time, and transport losses with different sensor technology choices. The studies illustrate how compromises driven by technology can impact the performance of the system in secondary ways, emphasizing the need for a systematic plan for building protection.

The building interior is divided into several active zones, and air samples are taken in each zone. Air samples from the tubing transport system are processed by a series of biological and chemical detectors using a tiered detection-response approach. A fast but lower-accuracy sensor in each active zone triggers a slower confirmatory sensor if a threat agent is detected. The primary response option for building protection is changing the state of the HVAC system to limit spread of the initially localized internal threat.

The metrics used in the Immune Building Program are the fraction of building exposed (FBE) and the fraction of occupants exposed (FOE). Both metrics make assumptions about the current HVAC state and occupancy of the building. In the preliminary evaluations of the protection system, FBE and FOE are estimated over time. FBE and FOE can be converted into numbers infected and likely incapacitation and mortality rates if a variety of assumptions are made.

Discussion

Because the Immune Building system was retrofitted to an existing building, challenges arose in coordinating the required changes in the existing HVAC system. The optimal solution was to have a control system for response in parallel with the control systems for normal operations. Because testing of the configuration at the Nord Hall test bed had not been completed at the time this report was written, the committee did not have sufficient information to evaluate the system or to evaluate the utility of the data to other deployments. A comparison of deployment choices made with other test beds and deployments suggest that Nord Hall represents a typical state-of-the-art system that is a compromise between limitations of current sensor technologies (in both performance and cost) and the development of a working system. A comprehensive evaluation of Nord Hall protection requires that the planned test of the technical performance of the system be completed and that integrated building protection systems that include operational responses be developed and tested. An important lesson learned is that test beds cannot be technology-only demonstrations; they must also demonstrate an integrated system that involves all components of an actual operating system including the response.

Pentagon

Description

The building protection system of the Pentagon represents the highest standard among the protection systems that the committee examined, and it is continuously being evaluated and improved. Although information on the building protection deployed at the Pentagon is limited, a general description and remarks can nevertheless be made in this report. The system uses all levels of protection (LPs as described in Chapter 3) and the choices of deployment corresponding to each level of protection.

- LP-1: The Pentagon has won numerous awards for providing a healthy working environment by minimizing natural air contaminants.
- LP-2: Many additional filtrations, including local air-washes, are deployed and combined with segmented protected spaces to provide optimized passive protection and to localize airborne threats.
- LP-3 and LP-4: The Pentagon uses a variety of sensor technologies in a tiered approach, both temporally and spatially separated, to provide fast low-regret response with longer-term confirmatory identification for treatment. The protection system also integrates remote external sensing with aerosol transport and dispersion, meteorological data, and airflow models to optimize response options. Similar integrated technologies are used within the building to supplement sensor information on the status of an event and to optimize response options. Visual monitoring systems are integrated into the sensor systems. Although certain functions are automated, the protection system obtains a high performance through regular training of personnel and evaluations of the systems.

Discussion

Two main observations about the protection system can be made. First, although the Pentagon represents a high-value asset where cost considerations are less important than other Department of Defense (DOD) facilities, the deployment represents the likely future paradigm of balancing and integrating all aspects of technology and operations to provide a robust, high-performance, and maintainable protection system. Second, because the Pentagon protection system is in a large and complex facility that captures many of the aspects of smaller buildings, use of the acquired data on performance and costs for modeling and to guide other deployments is highly desirable.

High-Asset Federal Building Deployment

The committee also considered other deployment across the federal complex to protect high-asset buildings, including the Joint Program Executive Office for Chemical and Biological Defense's Guardian Program and the Environmental Protection Agency's (EPA) Safe Building. Because of the breadth of the building types and the levels of protection required, only broad descriptions and observations can be made.

The types of buildings being protected represented a variety of existing and new buildings, and the following conclusions are drawn from the deployments:

- A case-by-case analysis is required.
- LP-1 and LP-2 options (no sensors) are the most broadly applicable options for most buildings given the current sensor technology and cost restrictions.
- Recommissioning and continual commissioning are essential to sustain performance of the LP-1 and LP-2 options.
- Testing (of air infiltration in particular) is essential.
- If the risk warrants, segmented internal spaces with cascading pressure zones maintained by simple control systems are deployed.
- If the risk warrants, separate air-handling units are used to isolate public, nonpublic, and safe areas.
- Models and simulations are useful to show areas of concern where insufficient data are available on airflow, air pathways, airtightness, and opening descriptions, for example.

The above guidelines and observations do not represent the gold standard of building protection but, rather, the affordable complex-wide options for building protection given the cost and limitations of current sensor technologies.

DESIGN AND SELECTION TOOLS

Building Protection Toolkit (Immune Building Program)

Within the Immune Building Program, an extended effort and substantial resources were focused on developing a multipurpose toolkit to support planning for building protection. To a lesser degree, the toolkit provides real-time response to determine dosages of occupants after an event. The BPTK integrates a collection of resources from many developers and covers the following:

- User input: architectural drawings, population data, scenarios involving different threat types, and external threat environments
- Conversion tools: creation of three-dimensional building representations, databases, and models of threat types and Immune Building technologies;
- Tools for assessing protective architectures: fast-running contaminant transport models—both indoor and outdoor; occupant mobility models for evacuation and gaming; and graphical interfaces for investigating multiple threat scenarios and Immune Building technologies, including cost estimations

The output of the toolkit captures the time-resolved history of FBE and FOE (metrics used in the Immune Building Program) as a function of cost. Use of the toolkit is aided by having predefined libraries of threat agents, filters, and sensors.

Discussion

Although the BPTK has not been fully developed or deployed, it is an attempt to provide a complete resource for building protection design. It integrates many of the component efforts around the country. Given the wide threat spectrum, variations in target buildings, and their complex interaction with other factors, a comprehensive analysis of the protection and cost options for a facility is difficult. Sophisticated tools such as BPTK have the potential to provide such analysis on a cost-effective and case-by-case basis and to allow generalization of data obtained from the few test beds and deployments.

The BPTK was developed as part of the Immune Building Program. Other government agencies have also supported the development of analysis and decision-making tools. Like the BPTK, these tools are new and have received only limited testing and application use.

Life-Cycle Cost Analysis Tool for Building Protection

The National Institute of Standards and Technology's Building and Fire Research Laboratory developed a tool for analyzing life-cycle cost with sponsorship from the EPA's Safe Buildings program to provide guidance to decision makers (of public facilities in particular) who are considering retrofitting their buildings to protect against biological and chemical attacks (NIST, 2006b). The Life-Cycle Cost Analysis Tool (LCAT) for building protection from biological and chemical airborne threats is based on economic tools that allow decision makers to consider options for protection components, installation, operation, and maintenance of a system (NIST, 2007). Use of the LCAT allows consistent comparison and contrast of the likelihood of different options to reach both protection and budget goals. The LCAT can be used to plan a building protection system and to evaluate its efficacy and cost, including unexpected expenses. LCAT is available publicly at *http://www2.bfrl.nist.gov/software/LCCchembio/index.htm*.

BPTK and LCAT are useful tools, but other design tools for building protection exist. For example, the Chemical-Biological Protection Tool (a tool for screening potential security upgrades) developed by the Technical Support Working Group and CONTAM PWC (a modified version of the publicly available CONTAM multizone modeling program) developed by the United Technologies Research Center have useful applications to building protection. Any plans for building protection design or the design of selection tools should consider as many available resources as possible prior to the design of building protection systems or the selection of tools.

OTHER RESOURCES

Security Design Criteria

Following the 1995 attack on the Alfred P. Murrah Federal Office Building in Oklahoma City, the U.S. Marshals Service was commissioned to perform a national study on vulnerabilities of federal buildings to terrorist attack. The report, *Vulnerability Assessment for Federal Facilities* (DOJ, 1995), was released in June 1995, and among its recommendations was the creation of a permanent Interagency Security Committee (ISC) by executive order (EO) to address physical security concerns of the federal government, including development of government-wide standards. ISC was established by EO 12977 in October 1995 and comprises 14 agencies. Currently under DHS, ISC has published and updated the *Security Design Criteria* (ISC, 2004a,b) for all nonmilitary, federally owned and leased properties. Although the security design criteria were primarily driven by considerations of blast mitigation, Chapter 5 of the report, "Mechanical Engineering," addresses some aspects of chemical, biological, and radiological threats. The ISC *Security Design Criteria* is a "for official use only" document. Some federal agencies have supplemented these security design criteria to cover specialized needs for such institutions as the National Institutes of Health and the Department of Veterans Affairs.

Unified Facilities Criteria

In 2003, DOD, through multiple uniformed services, developed the *Unified Facilities Criteria (UFC): DOD Minimum Antiterrorism Standards for Buildings* UFC 4-010-10 (DOD, 2003). Similar to the ISC security design criteria, the UFC criteria are directed primarily at blast mitigation. However, the U.S. Army Corps of Engineers also drafted *Protecting Buildings and Their Occupants from Airborne Hazards* in 2001 (U.S. Army Corps of Engineers, 2001). These documents are for unrestricted distribution.

Guidance Documents

After the terrorist attacks on September 11, 2001, guidance on building protection shifted not only from incidents that could be effectively controlled by security to incidents that require detailed strategy and planning of systems but also from being primarily provided by the government to government designers to being provided by a broad spectrum of key entities. Many guidance documents on how to reduce the impact of airborne biological and chemical attacks have been issued. In the case of airborne releases, the HVAC system can be an important weapon for thwarting or responding to an attack. Thus, these documents present a substantial reference library that can be used to guide building protection strategies that encompass risk management, physical security, protective technology, protective action, maintenance and commissioning to improve building security and response to attacks by using the HVAC system. The guidance documents include the following:

- *Building Security Through Design* (AIA, 2001)
- *DOD Minimum Antiterrorism Standards for Buildings* (DOD, 2003)
- *Design and O&M: Mass Notification Systems* (DOD, 2002)
- *Securing Buildings and Saving Energy: Opportunities in the Federal Sector* (Harris et al., 2002)
- *Addressing the Threat of Terrorism: Guidelines for Prevention and Response* (IFMA, 2002)
- *Protecting Buildings from a Biological or Chemical Attack: Actions to Take Before or During a Release* (LBNL, 2003)
- *National Air Filtration Association Position Statement on Bio-Terrorism* (NAFA, 2001)
- *Sheltering in Place as a Public Protective Action* (NICS, 2001)
- *Guidance for Protecting Building Environments from Airborne Chemical, Biological, or Radiological Attacks* (NIOSH, 2002)
- *Guidance for Filtration and Air-Cleaning Systems to Protect Building Environments from Airborne Chemical, Biological, or Radiological Attacks* (NIOSH, 2003)

- *Protecting Buildings and Their Occupants from Airborne Hazards* (USACE, 2001)

It would be wise to consult a variety of available resources to develop the best overall building protection strategy.

CONCLUSION

Information collected from test beds and current deployments is insufficient to provide comprehensive guidance on protection options for buildings across the DOD complex. Although many lessons were learned, the ability to extrapolate data and results that are specific to one facility to other facilities and situations is difficult to assess. Some observations can be made despite these limitations. (The role of test bed and decision support tools in a process for deployment of building protection is discussed in detail in Chapter 6.)

When the different groups of threat agents are considered (see Chapter 2), the group to which most buildings are most vulnerable—"cannot detect and cannot treat"—is not addressed by the more advanced technologies of LP-4. The only options to address these greatest vulnerabilities are the LP-1 and LP-2 options. Therefore, the building protection systems deployed in many high-asset federal buildings focus on LP-1 and LP-2 approaches.

The committee observed that some existing programs considered the initial costs of a building protection system and paid less attention to maintenance and operation costs, which have to be sustained by operational funds or some other continuous funding source. Both initial and life-cycle costs (that is, initial costs plus maintenance and operation costs) are higher for active than passive protection. The increase in cost limits the sustainability of active protection at present, except in the highest-asset facilities. The cost is likely to decrease and sustainability is likely to improve as the accuracy and reliability of sensor systems improve over time. Decision support tools such as BPTK will become important integrative tools for the design and implementation of building protection, particularly if these tools become repositories for performance data and costs of current and future deployments.

6

Deployment and Decision-Making Resources

Buildings are conceived and built in a linear process beginning with the definition by the building's owner of the needs the building is meant to fulfill. The process is generally known as the development of a project's functional program. For some new buildings, such as those owned by federal agencies and multinational corporations, this process might include a threat assessment and risk analysis (TARA). For buildings subject to the Interagency Security Committee Security Design Criteria, TARA is conducted by a multidisciplinary team as early in the process as feasible. Several agencies have issued guidance documents on conducting such assessments for building design and protection (NIOSH, 2002; ASHRAE, 2003; FEMA, 2003). Ideally, TARA is done before site acquisition because the choice of site location, access, and dimensions can affect physical security profoundly. For existing buildings, a vulnerability study is needed whereby the building is evaluated for its vulnerabilities to the threat and risk defined by the multidisciplinary team.

A functional program is developed by the building owner and key stakeholders who will occupy the building. From the functional program, an architectural space program is developed. The architectural space program establishes net and gross areas for the project including requirements related to the protection of building occupants from biological and chemical airborne threats as identified in the TARA. From the areas and owner-stated quality and performance objectives for systems, finishes, and other elements, a cost estimate can be developed using public and private cost databases for similar facilities. If the cost estimate exceeds available resources, the programs and quality objectives would be revised until the budget objectives are achievable.

Bidding documents including detailed drawings and specifications are then developed, and increasingly detailed cost estimates can be prepared. When the drawings and specifications are ready and the final cost estimate has been approved, bidding or another procurement method can begin. Typically, public work is subject to competitive bidding, but increasingly some agencies have used design-build as a procurement method. The design-build method usually involves preparation of preliminary drawings and specifications (the design development documents). These documents are used to solicit competitive bids for the completion of the design and construction as a package with a fixed lump sum or a guaranteed maximum price.

RISK ASSESSMENT AND RISK MANAGEMENT

The Concepts

Many fundamental concepts of human health risk assessment from chemical and biological hazards have been described in *Risk Assessment in the Federal Government: Managing the Process* (NRC, 1983). The basic paradigm developed in that report is shown in Figure 6-1, and it captures the two key components—risk assessment and risk management—that apply equally well to building protection situations.

Generally, human health risk assessments for biological and chemical threat agents include a risk assessment component that addresses the identification and characterization of agents involved (hazard or threat assessment), exposure potential for scenarios of interest (exposure or vulnerability assessment), and important uncertainties to fully characterize the resulting risk estimates (risk characterization or consequence assessment). The output of the risk assessment is used in the risk management process to identify and prioritize risk reduction strategies. Thus, risk management is the process of weighing alternatives and selecting the most appropriate actions that often integrates the results of risk assessment on human health risk with social, economic, and political concerns to reach a decision (NRC, 1983).

Implementation of a systematic approach for decision making for building protection requires input of various experts, including experts in medicine, health sciences, security and infrastructure protection, and building use and design. In addition to building protection, risk assessment and risk management have been applied to many other areas. Each area of application has fairly well-developed approaches that are optimized to its unique needs. Consequently, the areas of application might use somewhat different terminology and methodology. The following summary of risk assessment and risk management presents the highlights of what is judged to be common across different disciplines with potential applications to building protection.

Risk assessment provides an objective, often science-based, approach to compare risks. Risk assessments are inputs used by decision makers to deter-

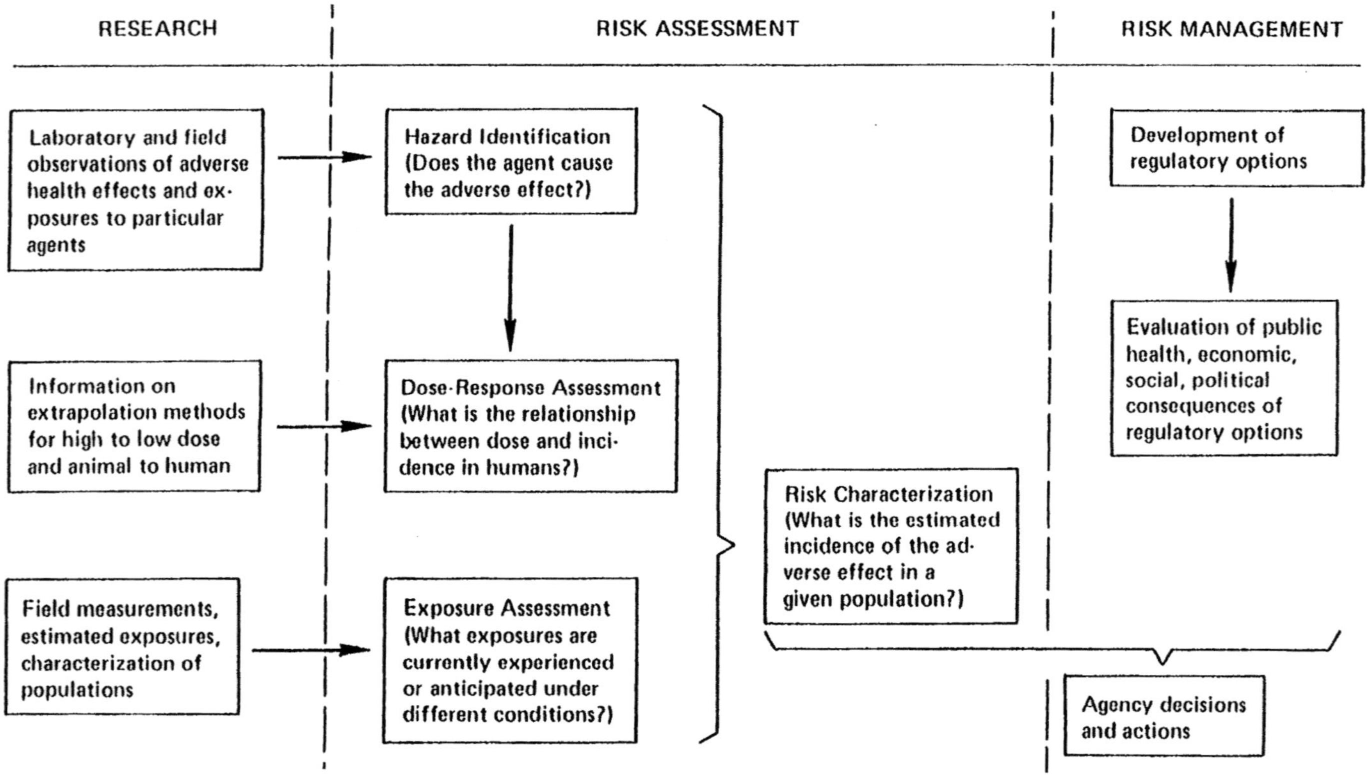

FIGURE 6-1 Elements of risk assessment and risk management of public health to biological and chemical and hazards (NRC, 1983).

mine risk management strategies. Risk management identifies the optimal course of action, taking into account conflicting objectives and uncertain supporting information.

Using the terminology of risk analyses for infrastructure protection, risk assessment is a process by which threats, vulnerabilities, and consequences are identified and used to inform decision making, including allocation of resources. All risk assessment and risk management approaches include common activities: (1) identification, characterization, and assessment of threat types; (2) assessment of the vulnerability of critical assets to threat types; (3) determination of the risk or specific consequence of attack or unintentional release; and (4) management of risks. Although the discussion below describes these activities separately, they are typically tightly coupled and interdependent. For example, ranking of threat types of concern depends on the determination of vulnerabilities, which in turn depends on the prioritization of mission goals, and so on. The information exchange among various components of risk assessment–risk management effort is also noted in the framework presented in the 1983 National Research Council (NRC) report.

Identification, Characterization, and Assessment of Threat Types. The characterization of threat type is required for a risk assessment. Important characteristics to consider include the physical, chemical, and health hazard characteristics of the threat agent; the intentional or unintentional nature of the threat type; the motivation for use of the agent; the triggers that might initiate an event; the method of delivery; and the trends seen from previous events (CRS, 2004). One of the greatest challenges in risk assessment for building protection is that the threat type is almost infinite in variety and complexity, but the risk assessment process still requires a characterization of the variety of threat types. Determination of the threat type is particularly challenging in the adversary-protector dynamic of building protection (NRC, 2007). The aim of the adversary is to seek and use only threat types of high consequence, whereas the aim of the protector is to remove vulnerabilities of high consequence. The adversary-protector dynamic is fundamentally different from that in other areas of risk assessment application such as environmental hazards or industrial safety where the variety of threat types is substantially less and more statistically determined. Because it is unrealistic to consider all imaginable threat types in a risk assessment, some metric is required to determine when the consideration of the threat types is comprehensive enough to cover most high-consequence threat types. An appropriate metric is evaluating the sensitivity of the conclusions of the risk assessment to the consideration of additional threat types (that is, ones that have not previously been considered in the analysis). If the conclusions are insensitive to the consideration of additional threat types or scenarios, then the threat-type characterization can be considered comprehensive. The additional challenge is that threat-type characterizations are sensitive to rapid changes in technologies and adversaries, so the characterization of threat types needs to be reassessed periodically.

Assessment of the Vulnerability of Critical Assets to Threat Types. Vulnerabilities specify the system components or operations that have a high sensitivity of the performance metric to a given threat type; for example, a given building is highly vulnerable to an indoor release of a given threat agent at a given location. Generally, vulnerabilities are dependent on all aspects of the protection systems—the physical, operational, technical, and organizational. Because all protection systems have vulnerabilities at some level, the greatest concern arises for vulnerabilities that result in severe consequences from a moderate threat type (a threshold or nonlinear response in the system). Also, a vulnerability that can be impacted by a wide variety of threat types, even though the consequence of the vulnerability might be less severe, should rank high as a vulnerability of concern. The importance of the number of threat types that affects vulnerability reinforces the introductory comment that the activities of risk assessment are tightly coupled and interdependent. Furthermore, the challenges cited above on the dynamic nature of threat-type characterization equally apply to the vulnerability assessment from the protector's perspective. Because of the wide variety of vulnerabilities and their dependence on mission and threat types, generalizations of vulnerabilities might be useful, as proposed in Figure 2-1, in focusing the development of new protection efforts across DOD facilities.

Determine the Risk or Specific Consequence of Attack or Unintentional Release. Risk can be seen as a discounted measure of consequence or an estimate of expected loss. It is based on what might or might not happen given the vulnerabilities, likelihood of an event, and importance of assets. The determination of risk can be viewed as the process that integrates the activities described above; it is the process that has the widest variety of methodologies across different disciplines that range from qualitative comparisons of perceived risk to quantitative methods with detailed causality networks and uncertainty quantification. In order to be sustainable in our complex and changing world, the method should be defensible (results clearly traceable to inputs and process) and adaptable (results can be updated quickly if new information is discovered). The risk assessment is a tool for providing the information needed to inform decisions across potentially conflicting missions and corresponding requirements. For example, significant consequences of an airborne release of biological and chemical threat agents include injury or death of occupants, disruption of key operations, and damage to or contamination of contents. The resulting impact from any of these consequences varies in severity depending on what is deemed critical, which needs to be outlined in the risk assessment.

Management of Risks. An overall risk reduction strategy can be achieved by characterizing the risk and benefit of each strategy as identified in the risk assessment. In building protection, risk reduction management would determine what level of protection and appropriate components are necessary for a specified protection objective. Depending on the building mission, different metrics of performance (for example, fraction of occupants exposed, fraction of building exposed) are weighted by their importance. The performance metrics establish the basis by which comparisons can be made across a variety of conflicting requirements. In most application areas, the common metric is often reduced to cost, including health consequences and loss of life. Because resources are always limited, the feasibility and cost of a given management strategy have to be assessed. Risks can be reduced by addressing any or all of the three components: the threat, vulnerabilities, or consequences. Once reasonable risk reduction strategies have been identified, the benefit of each strategy can then be characterized. As an example, while level of protection 3 (LP-3) strategy might be desirable for a particular facility, if limitations exist for installing or acquiring a particular component or if highly trained personnel are not available to maintain the system, it might be better to deploy LP-2 until resources to reach LP-3 can be acquired. In most risk management plans, costs play a significant role and are addressed in some form of a cost-benefit analysis. For building owners with a fixed budget, it is optimal to maximize the benefits for a given cost. There are two extremes for optimization: protect critical assets at the expense of others or protect all assets with marginal protection. Most owners, however, use strategies that are somewhere in between the two extremes. In all cases, cost-benefit analyses are important risk management tools used in setting priorities. (Cost consideration and cost-benefit analysis are discussed in further detail in the next two sections of this chapter.)

Guidance Documents for Building Protection

A number of private and governmental organizations, including the National Institute for Occupational Safety and Health (NIOSH), American Society of Heating, Refrigerating and Air-Conditioning Engineers, Inc. (ASHRAE), Federal Emergency Management Agency (FEMA), and Lawrence Berkeley National Laboratory (LBNL), recognize the need for guidance on protection of building occupants from terrorist attacks. They have issued documents intended to assist in the selection of security upgrades for managing risks from terrorist attacks (NIOSH, 2002; ASHRAE, 2003; FEMA, 2003; DOE, 2004). These documents provide specific guidance, such as upgrading filtration and relocating or securing outside air intakes. They also articulate the need for a rational multihazard assessment of risk and the application of a risk management process to evaluate possible courses of action. Such methods are necessary because the budget for protecting a building from natural and terrorist threats is finite, and thus the use of risk-based tools is necessary for priority setting.

The NIOSH guidance emphasizes the importance of understanding the building and establishing a list of its vulnerabilities (NIOSH, 2002). The Building Vulnerability Assessment and Mitigation Program (BVAMP) developed by LBNL (DOE, 2004) is an assessment tool that uses responses to a set of questions posed to building users to assemble an evaluation of the building and make recommendations for improving building safety.

The ASHRAE report outlines a four-step risk management process: (1) risk analysis, (2) risk treatment planning, (3) risk treatment plan implementation, and (4) reevaluating the plan after implementation and modifying it as needed. Risk analysis is itself a multistep process that involves (1) determining the organization's level of exposure, (2) identifying the risk, (3) estimating the probability of risk occurrence, (4) determining the value of the loss, and (5) ranking the risks and identifying the building's vulnerabilities. Each of the steps in risk analysis can be performed by a variety of methods, ranging from mainly heuristic to highly quantitative. The example assessment included in the ASHRAE report relies on subjective rather than analytical methods (ASHRAE, 2003). An updated edition of that AHSRAE report is being prepared (ASHRAE, forthcoming).

FEMA 426 provides a comprehensive discussion of terrorist threats to buildings and measures for mitigating them. It summarizes information found in many sources, including the NIOSH and ASHRAE documents mentioned above. The manual begins with a lengthy discussion of risk assessment (Figure 6-2), which provides the context for subsequent discussion of protective technology (FEMA, 2003).

In addition to this array of guidance, the Department of Homeland Security (DHS) is actively funding research centers (academic and national laboratories) to develop threat and vulnerability assessment approaches, including efforts geared to infrastructure protection. For example, DHS-funded research on assessment approach is conducted by the University of Southern California's Center for Risk and Economic Analysis of Terrorism Events, the University of Wisconsin-Madison, and other research groups. In 2005, DHS also conducted an extensive biological risk assessment covering many threat agents and many scenarios as directed by Homeland Security Presidential Directive 10 on *Biodefense for the 21st Century* (White House, 2004). These DHS-funded projects were presented at the Society for Risk Analysis Meeting in December 2006 (SRA, 2006) and were being reviewed by an NRC study (NRC, 2007) at the time this report was completed. The growing overall body of work highlights the need for a systematic risk assessment and risk management approach to assess building protection requirements. The specific methods and approaches for conducting such assessments are evolving. Current methods are highly variable and depend on the nature of the requirements being addressed. The DHS-funded projects and the evolving methods and approaches for conducting risk assessments are useful guidance and resources for the development and implementation of decision support tools in the context of a building protection program (NRC, 2007).

Because of the obvious variations in goals and objectives and in risk assess-

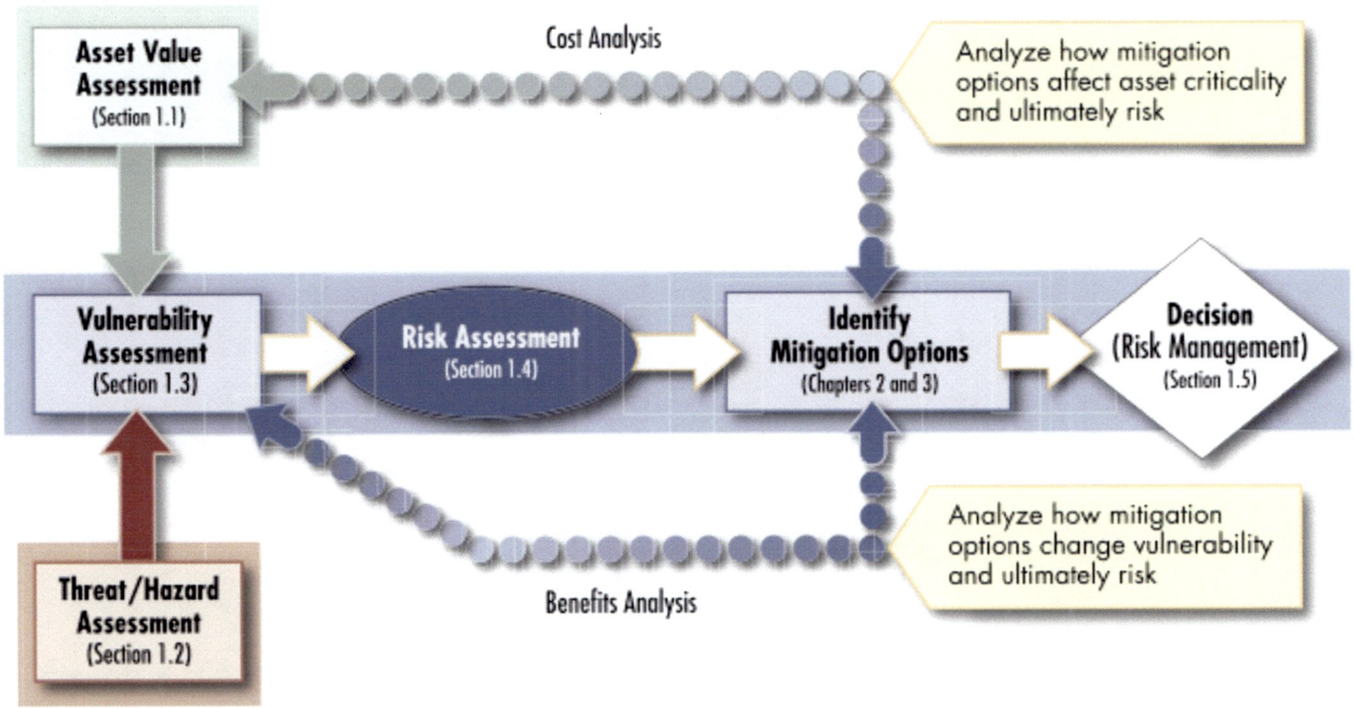

FIGURE 6-2 Risk assessment process model (FEMA, 2003).

ment and risk management approaches, specific methodologies or tools for risk assessment and risk management are not recommended here (some existing tools are described in Chapter 5 of this report).

COST CONSIDERATIONS

There are a number of components associated with the cost of building or retrofitting a facility for biological or chemical protection. Reliable costs cannot be readily calculated because construction and operating data are usually based on a history of similar buildings. As is the case with all construction, the actual costs associated with a project are known only after the facility is built. Furthermore, operation and maintenance costs are known only after a period of actual operation. The entire life-cycle cost[1] of a facility is estimated on the basis of the length of time the building is to be in operation. The shorter the building is expected to be in operation, the higher might the cost per year to operate become. Oftentimes, long-term use of a building might require extensive modernization; these costs are calculated on the basis of the history of similar buildings.

Costs of Construction

Costs of construction generally include construction materials, labor, and equipment. Construction costs are a subset of the actual total building cost, which includes permanent fixtures and equipment specified by the user. Material cost depends on the market cost of the construction materials and the quality of materials desired. The costs of equipment and labor depend on the market costs and a building productivity rate. A building to be built on a shorter schedule generally has higher costs for labor and equipment. Higher costs are attributable to increased work hours for a specific number of laborers or an increased number of laborers and to the increase in cost for use of equipment or use of more productive equipment. The major drivers in the costs of construction are presumed to be quality of and time for construction.

The costs of constructing or retrofitting a building for protection might or might not include the cost of specialty biological or chemical sensor equipment depending on the level of protection sought. If specialty equipment is not included, the construction costs of a building with some passive protection are related mostly to the quality of and time for construction. Without specialty equipment and with time of construction being equivalent, a passively protected facility is more expensive than a standard facility because the former requires a higher quality of construction to meet the expected performance of the building.

---

[1]All possible expenses incurred during the lifetime of a system. Life-cycle cost includes initial costs of implementation and all expenses incurred during the period of operation of equipment or a system.

If specialty equipment is included, the cost of biological and chemical sensors and the cost of high-quality construction are included in the construction costs of a protected building. However, these are only the initial costs that determine the capital investment required for protection. Life-cycle costs, which could be more substantial than initial costs, have to be considered.

Life-Cycle Costs

In addition to the initial costs of planning, design, and construction, periodic and preventive maintenance, repair, replacement of parts, and modernization incur additional costs during the life of a building. Comparing total costs or life-cycle costs is the most effective way to assess various alternatives to facility procurement. Future costs to repair or replace parts can only be estimated, but they are part of the budget consideration in a well-planned project. Fitting a building with a protection system would be fiscally unwise if the operation of the total system cannot be maintained. A well-functioning protection system involves a well-maintained building with minimal leakage and predictable airflow, well-maintained and calibrated sensors and response platforms if they are used, and periodic equipment replacement.

Estimating first costs of proposed buildings is best done from actual design documents—the more detailed these are, the better the estimate is. The more specialized and complex the building, the more important it is to have a detailed design as a basis for cost estimation. Pre-design cost budgeting is done from experience and using various public and commercial databases that provide costs per unit area or a similar metric. For unique building types or buildings designed for unusual requirements such as mitigating the effects of biological or chemical airborne threats, such databases do not exist or are not based on a large number of similar buildings. Therefore, useful total cost projections for mitigation of such attacks can probably be only done by pricing a range of hypothetical models in which enough design of systems and construction have been done to provide realistic conditions. Such analysis is beyond the scope of this study, but some important considerations should be noted.

- The first costs of design and construction vary even for the same building design. One variable is the familiarity of designers and constructors with the building type, its performance requirements, and technologies. For example, a design team and builder with experience in residential construction might not be able to design and construct a laboratory as economically or as satisfactorily as those with extensive experience in the field.
- The uniqueness of the design affects first costs. The repeated construction of a school or other building using "standardized" designs is likely to yield a lower first cost than construction of a one-of-a-kind building of the same function, especially if economies of large-quantity purchasing of materials can be

realized. Even buildings designed to be repetitive, such as some schools, housing, or utilitarian structures, have to be adapted to specific sites with varying topography and geology and to varying climatic conditions. These variations can affect first costs dramatically.

- The complexity of the design affects first costs. The greater the sophistication required of designers and constructors in technologies of design and building, the higher the first cost is likely to be. Design and construction of airtight buildings and sophisticated air distribution and filtration systems could add cost just because of the unfamiliarity with these requirements in current design and construction practice.
- A more subtle first-cost consideration might be in the process of procurement. The current local demand for construction labor and materials could drive the costs of both up or down. The rate of inflation applied to the design, procurement, and construction period and the cost to construction contractors of financing also affect the first cost of a building. Even the means of procurement such as design-bid-build versus design-build could produce different cost results for the same building design. Professional cost estimators with current knowledge of these conditions apply their judgment of the impact on each project estimated. Because these factors are time and location driven, generalized estimates are unreliable.
- Long-term life-cycle costs consist of those expenses for materials and labor necessary and desirable over the life of a building. Although some commercial and residential buildings are intended to last only for a few decades, other buildings of a public or institutional nature could last more than a century.
- The durability of a building and its components is affected by the quality of the original design, materials, and workmanship and the conditions under which the building exists. Climate—temperature, humidity, wind, and other elements—varies widely from one location to another, and so will the building's response. Repair and replacement costs correlate with the impact of age on the material and the severity of use and abuse by the building's occupants.
- Buildings require energy and personnel to operate. The design of a building directly affects the energy needs and costs. The quantity, skills, and time needed for operating personnel are also determined by the building design. Labor costs, including wages and benefits, are major components of life-cycle costs.
- All building systems and most materials require maintenance—from lubrication to refinishing and cleaning, to name a few typical categories. If a building owner has budgeted properly for replacement and repair, these costs can be managed over time. However, if maintenance, repair, and replacement have not been performed regularly and systematically, long-term costs can increase significantly. Many of the technologies needed to mitigate the effects of biological and chemical airborne threats in or near buildings depend on regular maintenance and verification of continued performance. These costs need to be taken into consideration in the planning and design of buildings. Inadequate long-term

operation and maintenance budgets can defeat the performance of the building and its protection system.

Scalability

An important consideration in designing building protection is the quantization of certain systems. Different levels of protection require different levels of technology to be installed. For example, a sophisticated biological detection capability must be installed for an LP-4 building (see LP-4 in Chapter 3) to provide timely and sensitive agent detection. The building has to be outfitted with the requisite technology to achieve LP-4. Setting up such capability is expensive and requires a major commitment in operations and operating budget. Irrespective of whether such capability is required for an entire building or for only a room within the building, the system and commitment level will be the same. Similar arguments can be made for many of the technologies discussed in this report (for example, filtration systems, detectors, triggers). In summary, monetary savings as a result of reducing the size of space to be protected has a threshold; when units are quantized, there is a minimum expenditure necessary to acquire the capability whether for a single room or for an entire building.

COST-BENEFIT ANALYSIS

A systematic approach to threat and consequence assessment for building protection supports decision making by prioritizing and optimizing risk management strategies that meet performance specifications for the building(s). However, the ability to meet performance specifications depends largely on the available resources. Cost-benefit analysis is a tool for optimizing performance and expenditure.[2] A cost-benefit analysis weighs the total expected costs against the total expected benefits of one or more actions in order to choose the best or most profitable option. In the case of a building protection system, total expected costs in the analysis include life-cycle costs of the system such as installation and maintenance of high-efficiency filters and expected benefits such as reduction of vulnerability to an attack, decrease in number of occupants exposed, and limited disruption to operation in the event of an attack. Cost-benefit analysis has played an important role in guiding decisions on human health protection (EPA, 2001), and methods for analysis have also been developed and used for homeland security applications (SRA, 2006).

The committee cannot present a general cost-benefit analysis for building

[2]A technique for project appraisal that weighs the total expected costs (including monetary costs and costs of risks) against the total expected benefits of one or more actions in order to choose the best or most profitable option. The technique is often referred to as CBA or as benefit-cost analysis in the United States.

protection because the costs and benefits of a protection system vary from building to building, but the application of cost-benefit concepts to building protection is discussed in this chapter. Strategies for optimizing the cost -benefit ratio for building protection can take many forms depending on the objectives. If the budget for protection is fixed, design tools (for example, the Immune Building Program's Building Protection Toolkit [BPTK]) could be used to identify the combination of design options to obtain maximum benefit, such as decrease in mortality, morbidity, or health-related costs. In contrast, if a specific health protection (or other performance metric) must be achieved, then the analysis focuses on identifying the combination design options that achieve the specified requirement for the least cost. In most cases, however, the amount of money that authorities are willing to spend on building protection depends on the specific protection to be achieved. In other words, a decision is made whether an incremental increase in funds to achieve a specific decrease in health risk is worthwhile. In that complex situation, the cost-benefit analysis focuses on optimizing the ratio of cost and benefit. Traditional cost-benefit approaches that use comparable metrics (usually monetary values) for both the design cost and the health benefits are useful.

Cost-benefit analyses (CBA) for each building protection system might have to be updated several times during the life cycle of a system. The first round of cost-benefit analysis is used to get concept approval to proceed with a more detailed one. A detailed cost-benefit analysis would then be conducted early in the design phase of the protection system. After the detailed analysis has been completed, the development and implementation plans might call for a prototype system or a pilot phase to test the costs and benefits on a small scale before the full system is implemented for all users. If a pilot phase is needed, a third version of the cost-benefit analysis would reflect revised costs and benefits and would be used to decide whether to proceed with full implementation of the system. The post-implementation review of a system might require an updated cost-benefit analysis to determine whether the expected benefits are being achieved and to decide whether the operation of the system should continue as implemented or be modified to achieve benefits that justify continued operation.

ROLE OF TEST BEDS AND DECISION SUPPORT TOOLS IN THE DEPLOYMENT PROCESS

An important objective of the Defense Advanced Research Projects Agency's (DARPA's) Immune Building Program is to provide a workable test bed to further enhance decision making regarding building protection options (DOD, 2006). These options reflect multiple levels of control and protection at a defined level of risk. Furthermore, many combinations of protection may be employed to achieve a desired mission requirement or specification (for example, see Chapter 3 for different levels of protection). Decision support tools can be applied at the planning or design stage to identify the optimum combination of protection features to achieve the specified mission goal and aid in decision making regarding tactics, responses, and remediation changes during and after an event. As noted above, the tools and approaches for threat, vulnerability, and consequence assessment are diverse. The subset of tools to provide comparative risk assessment and cost-benefit analyses are available and have been used in infrastructure protection applications to set priorities and to identify optimum resource allocation for buildings. Such tools are applied routinely in decision making related to implementing options for protecting human health. The DARPA Immune Building Program and related building protection efforts provide the opportunity to collect additional data to enhance threat, vulnerability, and consequence assessment tools. Also, test beds play an important role in data collection to support predictive modeling of threats, vulnerabilities, and consequences. Ideally, test bed facilities allow for testing of new technology options, such as sensors and filters, using a "plug-and-play" (that is, addition of a new device without reconfiguration) approach. In addition, such facilities provide the means to test the effect of alternative protection options on overall consequence, such as comparative risk. For example, the impact of physical security versus active ventilation systems versus operational tactical procedures on building protection can be compared. Test beds also allow information to be collected to fill existing data gaps, thereby improve the decision-making tools. Examples noted in presentations to the committee included data gaps regarding operations and management costs for building protection options and data on changes in building performance with aging. For these reasons, adequate test beds are needed to provide data for designing and refining threat, vulnerability, and assessment tools that facilitate decision making at the planning stage (building and design planning) and at the response and remediation stage.

A key criterion for assessing the value of maintaining a specific test bed facility is the extent to which the results generated by the test bed can be generalized or are applicable to other buildings of interest. Thus, the effectiveness and performance of various protective strategies (for example, impact of a new sensor on response time or fraction of building exposed) should apply to the buildings of interest.

Because none of the current test bed facilities fully represents all building designs or typical field conditions in most buildings, a series of virtual test beds representing each major class of buildings (see "Missions and Activities in Buildings" in Chapter 2) would fill data gaps and generalize lessons learned. Some technology-specific metrics are likely to be developed and refined, particularly for equipment evaluation (for example, sensor specificity and sensitivity, filtration efficiencies). Having a system for testing such technologies would also be an important feature of a building protection program. Nevertheless the ability to extrapolate the effectiveness and performance of a building protection system and all its components is needed. A critical evaluation of each test bed that exists is beyond the scope of this study. The committee is not aware of any sensitivity and uncertainty analysis that has been conducted to specifically assess the extent to which any single finding from a test bed can be used to make inferences about the broad array of buildings of interest. Because protection performance varies even among buildings that appear similar, quantitative data on system performance in similar facilities are needed to better predict the level of protection achieved by a potential design. For example, are two buildings similar if they have the same square footage, but one has multiple floors and one does not?

Because of the complexity of building protection and the difficulty of extrapolating data collected from test beds to other buildings, the need for predictive decision tools will likely be addressed by simulation modeling. For example, data collected across a variety of demonstration projects and deployments could be used to enhance modeling programs such as BPTK (Bryden, 2006). Enhancing these modeling tools is highly desirable because no single test bed can provide the data needed for accurate prediction of consequences from a biological or chemical weapon attack for the diverse array of facilities of interest. Although BPTK or other models could be useful, they might need further validation because only a few operational test beds have been modeled and subsequently tested. Maintaining test bed facilities would provide the opportunity for further testing and refining of interpolative and predictive models. A tool that incorporates a systemic approach—melding and hardening of concept of operations—is clearly needed to capture the full range of potential options for meeting a specified design goal. Although there is not a set of universal metrics and criteria for evaluation that applies to all building protection, a tool for decision making will document what metrics and criteria are most appropriate based on the goals and objectives of protection.

Software tools are also available to reduce the burden of deploying results of risk assessment and risk management across complex facilities. Those tools, such as CounterMeasures™ (Alion Science and Technology, 2007), are being deployed across the DOD complex to provide commanders with a resource to allocate limited funds to protect facilities from terrorist attacks (such as improvised explosive devices), but not specifically from biological or chemical attacks. The tools capture the four activities of risk assessment and risk management in a calculational database. They require surveys of vulnerabilities and protective actions in place as input and provide a cost-benefit analysis—on a dollar basis—of the building protection deployment options as an output. The tools can be used for a facility or across many facilities. Such tools could be extended for upgrading building protection from biological or chemical airborne threats and might be useful for planning complex-wide deployments of a building protection program.

CONCLUSION

Tools that can address the complexity of costs and benefits of building protection are available to assist stakeholders with different protection requirements and implementation budgets in designing and planning their systems. The risk and cost-benefit assessment tools used in conjunction with simulation tools lead to a decision-making process for the design and planning of building protection that is transparent (conclusions are defensible), comprehensive (addresses the complexity of the landscape), adaptable (can be modified quickly to address new information, such as detection technologies), and adjustable on the basis of needs (addresses the requirements of diverse stakeholders). Such a process would help to explain the choice of protection and justify its costs to diverse stakeholders.

7

Conclusions and Recommendations

Just as it is impossible to protect every citizen in every city from terrorist attack, protecting buildings operated by the Department of Defense (DOD) or within the civilian community will be imperfect, difficult, and costly. It is, however, feasible to systematically consider the options and implement those that are most cost-effective to achieve the defined goals and objectives of the facility. As the committee was conducting this study, many technical and behavioral issues were considered, but several stand out: all buildings are unique; buildings change and require maintenance and repair as they age; and detection and identification technologies have improved greatly in the last 20 years, but there are technical barriers that might not be overcome. The defense community needs to be cautious about seeking specific technical solutions too quickly without adequately considering simpler and often less expensive operational solutions. The complex and dynamic challenge of protecting humans and maintaining operational missions in buildings that might be at risk of an unknown attack with biological or chemical weapons can be addressed only if all the relevant factors are considered.

Recommendation 1: Clear and realistic building protection goals and objectives should be defined prior to deploying protection systems.

The Defense Threat Reduction Agency (DTRA) and other entities implementing building protection systems should clearly define the goals and objectives of building protection before and during the design phase or change of mission within the facility. In defining the goals and objectives, DTRA needs to identify and prioritize the critical resources that must be protected.

Recommendation 2: Building protection systems should be designed and implemented on a case-by-case basis for each structure to be protected.

The design and implementation of an appropriate and effective building protection system depends on many factors. These include the architecture, quality of construction, and condition of the building to be protected; the components to be used in the system (such as sensors and video monitoring); and the financial resources allocated for its design, implementation, and maintenance. To further complicate the matter, every building is unique because of the variations in its architecture and design, the materials used in and personnel who performed its construction, and wear and tear. All of these factors should be systematically considered before funds are committed to implementing building protection systems.

Recommendation 3: Life-cycle costs should be planned for prior to deploying building protection systems.

The complete life-cycle cost of a building protection system—including cost of planning, purchase, installation, maintenance, operation, and upgrade of all its components—should be considered prior to developing and implementing a building protection program. An effective building protection system requires proper integration of security technologies with building architecture and proper use of the system by building occupants. Integrating a protection system at the time of construction is typically less expensive and more efficient than retrofitting. Moreover, a building protection system will not be effective if it is not properly maintained, a significant consideration in life-cycle costs. A poorly maintained system quickly compromises the level of security expected of that system, leads to a false sense of protection, and could result in disruptive and expensive false alarms. For example, a high-level active protection system (level of protection 4 [LP-4], as described in Chapter 3) would only have performance equivalent to a passive protection system (LP-1 or LP-2) if the sensors are not maintained. Like passive filters, components of sensor systems have defined lifetimes and must be replaced periodically. They also need to be calibrated and tested for performance periodically. Therefore, a functional building protection system is not a one-time investment, but requires monetary resources for maintenance, repair, replacement, and upgrade of the system and its components. Finally, it is the exception within the federal government to budget for operation and maintenance costs, such as those that will be involved in building protection. Because these costs are likely to be higher than anticipated, advanced planning and budgeting are necessary to avoid loss of protection capability because it is seen by ultimately responsible local commanders as less important than the core mission.

The components to be used in a protection system are determined partly by the budget. Active protection systems tend to be more expensive than passive ones. State-of-the art sensor systems alone do not provide full protection; they need to be complemented by operational response plans. Although advances are being made in sensor technologies, more progress is necessary to determine how best to integrate them into systems and the implications for concepts of operation. Research would include systems studies to determine sensor locations and associated sensor requirements and appropriate concept of operations, along with validation of sensor performance in operational conditions (test beds or deployments). In addition, there is a need to consider the fragility and costs of operating and maintaining the sensor systems. Given the changing threats and life-cycle costs of advanced protection systems, it is possible that the most cost-effective and adaptive approach to protection for most buildings involves generic sensors that trigger only low-regret responses or even totally passive systems related to heating, ventilating, and air-conditioning (HVAC) without sensor technologies. Thus, inclusion of sensor technologies in a building protection program requires careful and systemic evaluation that weighs the costs and benefits of systems in a given potential spectrum of threats. Test beds provide an opportunity to collect data that can better inform the decision for future deployments.

Not all solutions to protection lie with detection and identification technology. A few solutions reside in building codes or regulations, though some are found in good design and construction practices. For example, building classifications exist on the basis of construction and ability to withstand fire, wind, seismic, and explosive events, but there is no building classification scheme for resistance to biological or chemical threats. Likewise, there are no uniform standards for establishing such classifications through standardized tests or metrics. Such standards and classifications could be developed by government or industry agencies, and DOD would play an active role in their development. Because of the limited protection offered by a modern "healthy building" and the better performance of advanced sensor systems in these environments, standards developed for healthy buildings will have positive impacts on building protection. Whether the protection system is active or passive, it needs to be evaluated periodically to ensure that protection goals are met.

Recommendation 4: Because goals and objectives for protection drive the choice of building protection system for each installation, metrics for a building protection system should be based on these same well-understood, clear goals and objectives.

The metrics for evaluating the effectiveness of building protection should be defined on the basis of the goals and objectives of protection. For example, if the goal is to maintain critical activities, metrics for evaluation might be continuity of operation and time to recover from an incident or to restore services. If the goal is to protect occupants, then metrics might include fraction of occupants exposed (FOE) or lives saved. Other criteria for evaluation that must be considered include life-cycle costs of the system, including the maintenance and operational cost of the system and its components.

Capturing data for some metrics is easier than for others. Thus, the ability of different metrics to accurately measure or estimate performance of a building protection system should also be considered in metric selection. For example, even though FOE might be a preferred metric to fraction of building exposed (FBE) for predicting adverse consequences to occupants, FBE might be selected in some cases where it is more feasible to estimate reliably. Based on these considerations, the committee cannot recommend any specific metric over another (see Chapter 4 for different possible metrics) but suggests that the selected metric be justified by the user as part of applying a systematic decision-making process for evaluating building protection.

Because biological and chemical attacks against buildings are rare events, performance evaluation of building protection systems in test beds will be necessary. The evaluation of sensors in a laboratory setting does not typically provide the sensor performance data (such as detection limit, false positive rate, and false negative rate) needed for designing building protection systems because of complex and variable backgrounds in buildings. Test beds, field studies, or deployments are valuable for evaluating the performance of individual sensors in building backgrounds and the performance of integrated sensor systems and building protection systems (with or without sensors). One advantage of test beds is that they can be configured and challenged in ways not possible in an operational facility.

For a test bed to be useful, an understanding of the extent to which technological performance results can be generalized to buildings of other types and for different missions is important. The extent to which technological performance results could be generalized varies. Because of the uniqueness of each building and the complexity of building protection systems, setting up test beds for each of the four levels of protection (described in Chapter 3) is impractical. However, an integrated test bed could be useful for testing aspects of building protection components and systems that could be applied to other buildings. Modeling and simulation methods could be developed and used in combination with experimental data to apply lessons learned from aspects of building protection components and systems within test beds and operational deployments to other buildings.

Having documentation and uniform protocols is important to obtain the most value from test beds and operational deployments. A continually evolving operational deployment such as the Pentagon could provide valuable real-world experience and data comparable to those obtained in test beds if all tests are well documented and standardized. For operational facilities or test facilities, establishment of uniform testing protocols to test effectiveness and validate protection systems would make their results useful to others. For example, information on degradation, maintenance, and operational and life-cycle costs (real and intangible) of building protection systems (including filters and HVAC, if applicable) could be collected in the test beds and existing deployments and used as points of reference for future cost analyses. Finally, data from a standardized but less comprehensive test protocol developed for use during commissioning of operational (non–test bed) facilities could be collected and fed back into virtual modeling programs to periodically upgrade the rigor and value of virtual testing and design resources.

Recommendation 5: Prior to implementation of a building protection program, the Department of Defense should establish a complete framework for building protection that guides decision making for each building to be protected. The decision-making framework should consider the following steps: (1) defining the objectives of building protection; (2) preparing a threat assessment; (3) establishing a risk assessment; (4) developing a case-by-case plan for building protection; (5) conducting a risk management analysis; and (6) analyzing costs and benefits using appropriate metrics and modeling and simulation tools as needed. The complexity of the steps in the framework and the time required for each step will depend upon the program and building protection objectives.

Designing and implementing an appropriate building protection system depends on the interactions of many factors—budget, objectives of protection, activities in the facility, location of the facility, and so on. Therefore, general principles apply, but a generic model for protection cannot fit all buildings. Rather, building protection has to be considered on a case-by-case basis. Thus, the committee can only provide guiding principles for designing and implementing a comprehensive decision-making framework for building protection that integrates risk assessment and risk management throughout the design, implementation, and deployment processes. Based on the current process of building protection reviewed above, the committee proposes some guiding principles:

- Define the goals of building protection (for example, maintenance of operations or protecting occupants). The goals play a role in determining the levels of protection sought. Metrics for evaluating performance of the protection system can then be determined based on the goals.
- Prepare a threat assessment. For each building, a threat assessment determines the possible threats and their likelihood (on the basis of current intelligence and vulnerabilities of the existing or planned facility). Because the threat type is uncertain, threat assessments are typically prepared in the following order: for the entire complex, for the facility type, for the location of the facility, for the mission or activities at the facility, and possibly for the current state of alert.
- Develop a risk or consequence assessment. The results of the threat and vulnerability assessment are used to prepare a risk or consequence assessment. The risk assessment establishes consequences for the various threats in the con-

text of vulnerabilities, and then ranks the possible threats and outcomes based on the requirements of the facility (for example, continuity of operations; limited personnel exposure). The risk assessment provides trade-offs in benefit for different levels of building protection and would capture uncertainties in the threat and effectiveness of the detection and response options.

- Conduct a risk management analysis. A risk management analysis is used to manage uncertainties in the effectiveness of different protection options. Analyze costs and benefits. The combination of risk assessment and estimated life-cycle costs provides a cost-benefit analysis of the protection options. Because a facility is part of a larger complex and the life-cycle cost of building protection is high, trade-offs across the complex must be done to consider retaining complex-wide function within a limited complex-wide budget.
- Develop a case-by-case plan for building protection (see Appendix E of DOD, 2005, for an example) that provides different options at different costs and then a building complex-wide analysis for allocation of limited resources.
- When construction (or retrofitting) begins, ensure throughout that the most up-to-date building plan is used, that all building modifications and plans are properly documented, and that the building is well constructed and maintained at all times.

Modeling and tools for simulation that can take into account different inputs for different buildings, protection systems, and costs need to be tested and validated before they become cost-effective resources for designing and identifying gaps in building protection. Building protection has not been tested extensively for efficiency and efficacy under a range of scenarios. Furthermore, the actual risk of biological and chemical attack is unknown. Thus, the effectiveness of current concepts of building protection accomplishing a mission is uncertain. Although models have been developed to assess the impact of threat types, they could be strengthened with broad application, as well as testing and feedback, because existing models do not cover many scenarios. Modeling and simulation tools used to trace the impact of threat agent releases can be effective only if they are developed based on the goals of building protection set forth through a complete decision-making process.

Although the above principles will apply to many buildings, deployment of building protection is dominated by a case-by-case analysis and implementation because of the uniqueness of each building, its mission, and its location and the current threat. Therefore, the above approach should be tailored to match the current perceived threats, known vulnerabilities, and the development of likely detection technologies in order to provide a balanced evolutionary path from the current state over the next 10–15 years.

Decision support tools can be used at the planning or design stage to identify the optimum combination of controls to achieve the specified mission goal and can aid in operational decision making regarding tactics, responses, and remediation changes during and after an event. Comparative risk assessment and cost-benefit analysis approaches can aid in all stages of decision making.

The combined decision support system for building planning would do the following:

- Capture the possible conflicting requirements of a variety of deployments (for example, protection of people versus property);
- Compensate for uncertainty in input data and performance achieved;
- Resolve conflict in expert input or data available;
- Possess the ability to determine knowledge or operational gaps in input data or operational approaches, respectively; and
- Provide a defensible approach where conclusions can be matched to data and process, and can be adapted to change as threat, test, and operational data become available.

Once a framework for building protection is established and validated by deployments, it can be used to develop standardized building protection responses that capture "best practices" from comprehensive considerations of threat types and cost-benefit analysis across many building types. The development of standardized building protection responses does not infer that consideration of building protection on a case-by-case basis is unnecessary. Rather, it recognizes that as experience is gained in building protection, it might become apparent that some classes of building types and uses might require similar deployments for protection.

Recommendation 6: Building protection should be designed to accommodate changing building conditions, emerging threats, and changing technology. Both the deployed building protection and the framework for deploying the building protection (proposed in Recommendation 5) should be reviewed periodically for sustained performance in light of changing resources and threats.

Plans should include having a building's protective design tested and reevaluated for performance, and revised or replaced in order to respond to changing parameters and needs. The spectrum of threats is ever changing so that the risk analysis for each building might be unique and changing as well. Buildings also evolve over time through aging, wear and tear of use, and the effects of climate. A lot of attention and funding have been focused on sensor and detection technologies. Human factors and concepts of operations have to be considered as well. New buildings could have an infrastructure (for example, power, communications, space) that allows for installation of new protection components as they become available. Although building protection has to be developed on a case-by-case basis, generic principles can be applied to many buildings and situations. The establishment of a systematic process that weighs the different options should guide decision making. Such a process would help DTRA and other agencies to design appropriate protection systems that optimize protection within the physical limitations of the building, technological limitations of the components (such as HVAC, filters, and sensors), and financial constraints. Developing the process that takes into account all the input for design to optimize output will incur additional costs up front, but the committee believes that the utility and ultimate cost savings of the approach will well justify this initial cost increase. To sustain the performance of a deployed building protection system or the framework to support its deployment, deployed building protection systems and their corresponding deployment processes should be reviewed periodically to assess whether they align with changing building conditions, building design methods, sensor and protection technologies, and threat types.

References

AIA (American Institute of Architects). 2001. Building Security Through Design. Washington, D.C.

Alion Science and Technology. 2007. CounterMeasures™. Accessed on May 13, 2007, at http://www.countermeasures.com/.

Allen, C., W. Jackson, L. Saubier, and C. Powell. 2006. Smart Building, Volume 1—Executive Summary, Report DTRA-TR-03-41-V1, Defense and Threat Reduction Agency, Fort Belvoir, Va.

American Lung Association. 2007. Factors to Consider Before Using/Buying an Air Cleaner. Accessed on February 1, 2007, at http://www.lungusa.org/site/pp.asp?c=dvLUK9O0E&b=39310.

ASHRAE (American Society of Heating, Refrigerating, and Air-Conditioning Engineers). 1992. ASHRAE Standard 52.1: Gravimetric and Dust-Spot Procedures for Testing Air-Cleaning Devices Used in General Ventilation for Removing Particulate Matter. Atlanta, Ga.

ASHRAE. 1999. ASHRAE Standard 52.2: Method of Testing General Ventilation Air-Cleaning Devices for Removal Efficiency by Particle Size. Atlanta, Ga.

ASHRAE. 2001. ASHRAE Handbook—Fundamentals. Atlanta, Ga.

ASHRAE. 2002. Risk Management Guidance for Health and Safety Under Extraordinary Incidents. Atlanta, Ga.

ASHRAE. 2003. Risk Management Guidance for Health, Safety and Environmental Security Under Extraordinary Incidents. Atlanta, Ga.

ASHRAE. 2004a. ANSI/ASHRAE Standard 62.1-2004 Ventilation for Acceptable Indoor Air Quality. Atlanta, Ga.

ASHRAE. 2004b. HVAC Systems and Equipment, SI Edition. Atlanta, Ga.

ASHRAE. Forthcoming. ASHRAE Guideline 29P, Guideline for Risk Management of Public Health and Safety in Buildings. Atlanta, Ga.

ASTM (American Society for Testing and Materials). 1998. E964 Standard Practice for Measuring Benefit-to-Cost and Savings-to-Investment Ratios for Buildings and Building Systems. West Conshohocken, Pa.

ASTM. 1999. E917-99 Standard Practice for Measuring Life-Cycle Costs of Buildings and Building Systems. West Conshohocken, Pa.

Bauman, F. 2003. Underfloor Air Distribution (UFAD) Design Guide. Atlanta, Ga.

Braun, R.H. 1986. Problem and solution to plugging of a finned-tube cooling coil in an air handler. *ASHRAE Transactions* 92:385-389.

Bryden, W. 2006. Building Protection Toolkit. Accessed on December 14, 2006, at http://www.darpa.mil/sto/chembio/pdf/BPTK.pdf.

Campbell, L., S. Talmont, R. Relyea, R. Cain, A. Policastro, W. Dunn, G. Lurie, and D. Brown. 2004. Integration of the PROTECT Chemical Detection System into WMATA Operations. In Proceedings of the Rail Transit Conference, Miami, Fl.: American Public Transportation Association.

CRS (Congressional Research Service). 2004. Risk Management and Critical Infrastructure: Assessing, Integrating, and Managing Threats, Vulnerabilities, and Consequences. RL32561. Washington, D.C.

DHHS (Department of Health and Human Services). 1984. Biosafety in Microbiological and Biomedical Laboratories (BMBL), 1st Edition. Washington, D.C.: U.S. Government Printing Office.

DHHS. 2000. Chemical and Biological Respiratory Protection Workshop Report, DHHS (NIOSH) Publication No. 2000-122. Accessed on March 26, 2007, at http://www.cdc.gov/niosh/2000-122.html.

DHHS. 2007. Biosafety in Microbiological and Biomedical Laboratories (BMBL), 5th Edition. Accessed on April 19, 2007, at http://www.cdc.gov/od/ohs/biosfty/bmbl5/bmbl5toc.htm.

DHS (Department of Homeland Security). 2006. BioWatch: Early Detection, Early Response. Accessed on December 12, 2006, at http://www.milnet.com/wh/DoHS/BioWatchFactSheetFINAL.pdf.

DOD (Department of Defense). 2002. Design and O&M: Mass Notification Systems. UFC 4-021-01. Washington, D.C.

DOD. 2003. Unified Facilities Criteria (UFC). DOD Minimum Antiterrorism Standards for Buildings. UFC 4-010-01. Washington, D.C.

DOD. 2005. Security Engineering: Procedures for Designing Airborne Chemical, Biological, and Radiological Protection for Buildings. UFC 4-024-01. Washington, D.C.

DOE (Department of Energy). 2004. Building Vulnerability Assessment & Mitigation Program (BVAMP). Accessed on January 9, 2007, at http://securebuildings.lbl.gov/BVAMP/BVAMP_brochure.pdf.

DOJ (Department of Justice). 1995. Vulnerability Assessment of Federal Buildings. Washington, D.C.

DOJ. 2000. Guide for the Selection of Chemical Agent and Toxic Industrial Material Detection Equipment for Emergency First Responders. Accessed on April 24, 2007, at http://www.ncjrs.gov/pdffiles1/nij/184449.pdf.

EPA (Environmental Protection Agency). 2001. Human Health Metrics for Environmental Decision Support Tools: Lessons for Health Economics and Decision Analysis. EPA /600/R-01/104. Cincinnati, Ohio.

EPA. 2006. Research Highlights 6/26/06. HVAC In-duct UV Light Systems. Accessed on February 28, 2007, at http://www.epa.gov/nhsrc/news/news062606.html.

FEMA (Federal Emergency Management Agency). 2003. Risk Reference Manual to Mitigate Potential Terrorist Attacks Against Buildings, Publication No. FEMA-426. Washington, D.C.: U.S. Government Printing Office.

Fisk, W.J. 2000a. Health and productivity gains from better indoor environments and their relationship with building energy efficiency. *Annual Review of Energy and the Environment* 25:537-566.

Fisk, W.J. 2000b. Estimates of potential nationwide productivity and health benefits from better indoor environments: An update. In Indoor Air Quality Handbook, J.D. Spengler, J.M. Samet, and J.F. McCarthy, eds. New York: McGraw-Hill Professional.

GAO (Government Accountability Office). 2002. Building Security, Security Responsibilities for Federally Owned and Leased Facilities. GAO-03-8. Washington, D.C.

Harris, J., W. Tschudi, and B. Dyer. 2002. Securing Buildings and Saving Energy: Opportunities in the federal sector. In The Austin Papers: Best of the 2002 International Green Building Conference. Brattleboro, Vt.

Heyman, D.L. 2005. Control of Communicable Diseases Manual, 18th Edition. Washington, D.C.: American Public Health Association.

Hinds, W.C. 1999. Aerosol Technology: Properties, Behavior, and Measurement of Airborne Particles, 2nd Edition. New York: John Wiley and Sons.

Hitchcock, P.J., M. Mair, T.V. Inglesby, J. Gross, D.A. Henderson, T. O'Toole, J. Ahern-Seronde, W.P. Bahnfleth, T. Brennan, H.E. Burroughs, C. Davidson, W. Delp, D.S. Ensor, R. Gomory, P. Olsiewski, J.M. Samet, W.M. Smith, A.J. Streiffel, R. H. White, and J.E. Woods. 2006. Improving performance of HVAC systems to reduce exposure to aerosolized infectious agents in buildings: Recommendations to reduce risks posed by biological attacks. *Biosecurity and Bioterrorism: Biodefense Strategy, Practice, and Science* 4:1-14.

IES (Institute of Environmental Sciences). 1992 IES-RP-CC007.1: Testing ULPA Filters (Recommended Practice). Mount Prospect, Il.

IFMA (International Facility Management Association). 2002. Addressing the Threat of Terrorism: Guidelines for Prevention and Response. Houston, Tx.

IOM (Institute of Medicine). 1999. Chemical and Biological Terrorism: Research and Development to Improve Civilian Medical Response, pp. 41-42. Washington, D.C.: National Academy Press.

ISC (Interagency Security Committee). 2004a. Security Design Criteria for New Federal Office Buildings and Major Modernization Projects: Part I. For Official Use Only.

ISC. 2004b. ISC Security Design Criteria for New Federal Office Buildings and Major Modernization Projects: Part II: Tables, Design Tactics, and Additional Risk Guidelines. For Official Use Only.

LBNL (Lawrence Berkeley National Laboratory). 2003. Protecting Buildings from a Biological or Chemical Attack: Actions to Take Before or During a Release. LBNL/PUB-51959. Berkeley, Ca.

Martin, S.B., B.R. Beamer, and E.S. Moyer. 2006. Evaluation of a high-efficiency, filter-bank system. *Journal of Occupational and Environmental Hygiene* 3:204-213.

Merrill, L., J. Dunbar, J. Richardson, and C.R. Kuske. 2006. Composition of *Bacillus* species in aerosols from 11 U.S. cities. *Journal of Forensic Science* 51:559-565.

NAFA (National Air Filtration Association). 2001. NAFA Position Statement on Bio-Terrorism. Washington, D.C.

NAFA. 2004. Types of HEPA filters. *Air Media* (winter).

NFPA (National Fire Protection Association). 2001. NFPA 1994: Standard on Protective Ensembles for Chemical/Biological Terrorism Incidents, Quincy, Ma.

NICS (National Institute for Chemical Studies). 2001. Sheltering in Place as a Public Protective Action. Charleston, Wv.

NIH (National Institutes of Health). 1999. Cost-Benefit Analysis Guide for NIH IT Projects. Accessed on February 7, 2007, at http://irm.cit.nih.gov/itmra/cbaguide.html#3.

NIOSH (National Institute for Occupational Safety and Health). 2002. Guidance for Protecting Building Environments from Airborne Chemical, Biological, or Radiological Attacks. DHHS (NIOSH), Publication No. 2002-139. Cincinnati, Ohio.

NIOSH. 2003. Guidance for Filtration and Air-Cleaning Systems to Protect Building Environments from Airborne Chemical, Biological, or Radiological Attacks. DHHS (NIOSH), Publication No. 2003-136. Cincinnati, Ohio.

NIOSH. 2004. NIOSH Respirator Selection Logic 2004. DHHS (NIOSH), Publication No. 2005-100. Cincinnati, Ohio.

NIST (National Institute of Standards and Technology). 2006a. NIST Multizone Modeling Website. Accessed on January 8, 2006, at http://www.bfrl.nist.gov/IAQanalysis/index.htm.

NIST. 2006b. Life-Cycle Cost Analysis Tool for Chem/Bio Protection of Buildings. Accessed on March 15, 2007, at http://www2.bfrl.nist.gov/software/LCCchembio/index.htm.

NIST. 2007. Building Retrofits for Increased Protection Against Chemical and Biological Airborne Releases. NISTIR 7379. Gaithersburg, Md.

NRC (National Research Council). 1983. Risk Assessment in the Federal Government: Managing the Process. Washington, D.C.: National Academy Press.

NRC. 2000. Acute Exposure Guideline Levels for Selected Airborne Chemicals: Volume 1. Washington, D.C.: National Academy Press.

NRC. 2002. Acute Exposure Guideline Levels for Selected Airborne Chemicals: Volume 2. Washington, D.C.: National Academy Press.

NRC. 2003a. Testing and Evaluation of Standoff Chemical Agent Detectors. Washington, D.C.: The National Academies Press.

NRC. 2003b. Acute Exposure Guideline Levels for Selected Airborne Chemicals: Volume 3. Washington, D.C.: The National Academies Press.

NRC. 2004. Acute Exposure Guideline Levels for Selected Airborne Chemicals: Volume 4. Washington, D.C.: The National Academies Press.

NRC. 2005a. Reopening Public Facilities After a Biological Attack: A Decision Making Framework. Washington, D.C.: The National Academies Press.

NRC. 2005b. Sensor Systems for Biological Agent Attacks: Protecting Buildings and Military Bases. Washington, D.C.: The National Academies Press.

NRC. 2005c. Monitoring at Chemical Agent Disposal Facilities. Washington, D.C.: The National Academies Press.

NRC. 2006. Acute Exposure Guideline Levels for Selected Airborne Chemicals: Volume 5. Washington, D.C.: The National Academies Press.

NRC. 2007. Interim Report on Methodological Improvements to the Department of Homeland Security's Biological Agent Risk Analysis. Washington, D.C.: National Academies Press.

OSHA (Occupational Health and Safety Administration). 1999. Personal Protective Equipment: Respiratory Protection. Occupational Health and Safety Standards 1910.134. Washington, D.C.

Ottney, T.C. 1993. Particle management for HVAC systems. *ASHRAE Journal* 35:26.

Richmond, J.Y., and R.W. McKinney. 1999. Biosafety in Microbiological and Biomedical Laboratories, 4th Edition, HHS Publication Number (CDC) 93-8395. Washington, D.C.: U.S. Government Printing Office.

Seppänen, O., and W. Fisk. 2006. Some quantitative relations between indoor environmental quality and work performance or health. *HVAC&R Research* 12:957-973.

SRA (Society for Risk Analysis). 2006. Society for Risk Analysis Annual Meeting 2006. Accessed on January 10, 2007, at http://birenheide.com/sra/2006AM/program/sessionlist.php3.

Stetzenbach, L.D. 2002. Introduction to aerobiology. Pp. 801-813 in Manual of Environmental Microbiology, 2nd edition, C.J. Hurst, R.L. Crawford, G. Knudsen, M. McInerney, and L.D. Stetzenbach, ed. Washington, D.C.: ASM Press.

USACE (U.S. Army Corps of Engineers). 1999. Engineering and Design—Design of Collective Protection Shelters to Resist Chemical, Biological, and Radiological (CBR) Agents. Washington, D.C.

USACE. 2001. Protecting Buildings and Their Occupants from Airborne Hazards. TI 853-01. Washington, D.C.

Vogt, F. 2006. Trends in remote spectroscopic sensing and imaging—Experimental techniques and chemometric concepts. *Current Analytical Chemistry* 2(2):107-127.

Walter, K. 2006. An Action Plan to Reopen a Contaminated Airport. Science and Technology Review, December 2006. Accessed on January 5, 2007, at http://www.llnl.gov/str/Dec06/pdfs/12_06.2.pdf.

Walton, G.N., and W.S. Dols. 2006. CONTAM 2.4 User Guide and Program Documentation (NISTIR 7251). Washington, D.C.: National Institute of Standards and Technology.

White House. 1992. Guidelines and Discount Rates for Benefit-Cost Analysis of Federal Programs. Office of Management and Budget Circular No. A-94 revised. Accessed on February 7, 2007, at http://www.whitehouse.gov/omb/circulars/a094/a094.html.

White House. 2004. Homeland Security Presidential Directive 10: Biodefense for the 21st Century. Accessed on January 29, 2007, at http://www.whitehouse.gov/homeland/20040430.html.

Appendixes

A

Statement of Task

At the request of the Defense Threat Reduction Agency (DTRA), the National Academies shall conduct a study to assist in DTRA's capacity to plan, design, construct, and operate future chemical and biological resistant facilities for the Department of Defense (DOD). The study shall give consideration to prior studies from at least the last five years; current operational uses of related systems; and current science and technology development efforts intended to improve upon existing building protection systems. It shall consider work from both defense and civilian sources. At a minimum the Academy shall address the following questions:

- What metrics of performance are relevant to evaluate existing studies and to use existing facilities as effective test beds for validating tools, testing systems, and facilitating system technology transfer? Where a metric is not relevant to all situations, identify and discuss its appropriate application. Discuss situational use of a combination of all relevant metrics where appropriate.
- What terms and definitions are required—for example, Tier 1 detector, trigger, high impact response, confirmatory test, etc.—to allow communication and comparison between programs?
- Consider the current protocols and setup of existing systems in use, including both DOD and non-DOD efforts. What are the general features of existing test bed facilities? Are there significant features in common? Do existing facilities differ in significant ways, and how can these differences be exploited to forward our understanding of building protection?

- What collective principles can be derived from current building protection efforts? How can information gained from a test-bed facility be extrapolated to operational buildings with completely different designs?
- What is the cost-benefit of internal building monitoring? Suggest a tiered approach with varying levels of detection and protection capability, comparing the relative cost-benefit between tiers. Perform this assessment for both new building construction and building retrofit, and correlate to an appropriate metric (lives saved/fraction building exposed).
- Compare and discuss the relative risks of the possible tiers in a tiered approach to chemical and biological protection efforts, from a baseline of no protection efforts up to and including a fully protected building. Consider risks associated with building retrofitting, extrapolating test data to buildings differing from test bed buildings, possible system degradation over time, etc.

B

Committee Member Biographies

David R. Franz *(co-chair)* is chief biological scientist at the Midwest Research Institute and director of the National Agricultural Biosecurity Center at Kansas State University. He served in the U.S. Army Medical Research and Materiel Command for 23 of his 27 years on active duty. Dr. Franz has served as deputy commander and commander of the U.S. Army Medical Research Institute of Infectious Diseases (USAMRIID) and as deputy commander of the U.S. Army Medical Research and Materiel Command. Before joining the command, he served as group veterinarian for the 10th Special Forces Group (Airborne). Dr. Franz was technical editor of the *Textbook of Military Medicine on Chemical and Biological Defense* and has been an invited speaker at many nationally and internationally recognized organizations. He served on the National Research Council (NRC) Committee on Biological Threats to Agricultural Plants and Animals. He is serving on the NRC Committee on Genomics Databases for Bioterrorism Threat Agents and on the Committee to Review Research Proposals from Former Soviet Biological Weapons Institutes, which he chairs. Dr. Franz holds a DVM from Kansas State University and a Ph.D. in physiology from Baylor College of Medicine.

Norman L. Johnson *(co-chair)* is chief scientist at Referentia Systems, a small, minority-owned business that develops advanced technology solutions to complex problems in the areas of defense and homeland security. He received his B.S. from the University of California, Davis and his Ph.D. from University of Wisconsin, Madison. Dr. Johnson is on leave of absence from Los Alamos National Laboratory where he served for 25 years, most recently as Deputy Group Leader of the Theoretical Biology and Biophysics Group in the Theoretical Division.

Before this, he was deputy program manager for three years for the Biological Threat Reduction Program Office, under Dr. I. Gary Resnick, and guided the development and execution of a $40 million program in all areas of biothreats, from genomics to sensor systems to system modeling to operations. As a project manger, he oversaw projects that were challenging and often considered to be in the "too hard to do" box. The key to success was enabling diverse teams to break limiting barriers and discover synergistic advantages of diverse contributions. His published research covers multiphase flows, inertially confined fusion, combustion modeling, self-organizing knowledge creation, diversity in collective systems, and developmental theories of evolution. His current areas of interest are biodefense, epidemiology—particularly pandemic influenza, and modeling the dynamics of social collectives and social identity.

William P. Bahnfleth is professor of architectural engineering and the founding director of the Indoor Environment Center (IEC) at the Pennsylvania State University. He has nearly 25 years of experience as a design engineer, researcher, and educator in the building mechanical systems field. He teaches and conducts research on systems for controlling indoor air quality and efficient utilization of energy in building heating, ventilating, and air-conditioning (HVAC) systems. Current areas of investigation include thermal energy storage, ultraviolet germicidal irradiation for the control of bioaerosols, demand-controlled ventilation, and design of HVAC systems to mitigate the effects of chemical and biological releases. As a consultant, he has assisted in the design of more than a dozen thermal storage systems in the United States, Canada, Australia, and Saudi Arabia. Dr. Bahnfleth served on the NRC Committee on Safe Buildings Program.

Cynthia Bruckner-Lea currently manages the Chemical and Biological Sciences Group at Pacific Northwest National Laboratory (PNNL), which is focused on chemical and biological detection and forensics research and development. She received a B.S. in chemical engineering from the University of California, Davis, and a Ph.D. in bioengineering from the University of Utah. Dr. Bruckner-Lea has developed several bioanalytical research programs for environmental monitoring and medical applications. For example, she led interdisciplinary research teams in developing automated pathogen detection systems based on nucleic acid analysis using planar microarrays and bead suspension arrays, a project team focused on the development of nanoparticle labels and methods for rapid antibody-based pathogen detection, and a multilaboratory team developing a broad-spectrum point biodetection system. She served as the chair of the Sensor Division of the Electrochemical Society from 2002 to 2004. She often organizes symposia and is an invited speaker at many sensor symposia. Dr. Bruckner-Lea served on the NRC Committee on Materials and Manufacturing Processes for Advanced Sensors.

Steven B. Buchsbaum is currently a senior program officer in the Global Health Technologies program at the Bill & Melinda Gates Foundation. His areas of focus at the foundation include statistical and modeling issues, vaccine delivery technologies, diagnostic platform technologies, technologies for etiological surveillance, and tuberculosis vaccine and drug discovery. Prior to this position, he was a program manager for both the Defense Advanced Research Projects Agency (DARPA) and the Homeland Security Advanced Research Projects Agency and had direct contact with the DARPA Immune Building Program. He is an engineer by training with a strong background in sensing and detection as well. Dr. Buchsbaum received an M.S. and Ph.D. in physics, as well as his MP.I.A. in international technology management, from the University of California, San Diego. He earned his B.A. from Hamilton College in New York.

Sheldon K. Friedlander *(deceased)* was a Parsons Professor of Chemical Engineering and director of the Nanoparticle Technology/Air Quality Engineering Lab at the University of California, Los Angeles. His research group works on aerosol engineering and the science and technology of fine particles in gases, with applications to air pollution and advanced materials. The synthesis of fine particles in narrow size ranges with controlled crystalline properties is an emerging technology with important industrial applications. Such particles when formed under uncontrolled conditions and emitted to the atmosphere may pose a threat to public health. His students have become familiar with (and taken jobs in) both the air pollution and the advanced materials fields of application. Dr. Friedlander was elected to the National Academy of Engineering in 1975 and served on different NRC committees.

Murray Hamlet is a retired chief of the Research Support Division in the U.S. Army Research Institute of Environmental Medicine (USARIEM) in Natick, Massachusetts. Prior to that, he served as director of the Cold Research Division, USARIEM, from 1972 to 1989. Long interested in threat, vulnerability, and forensics, Dr. Hamlet has conducted threat analyses of buildings in greater Boston against biological, chemical, and explosive threats. He inspects the HVAC systems; power and communications inputs; the gas, water, and sewage systems; the delivery and waste disposal procedures; and the surrounding streets and buildings. He also documents vulnerable points and practices, and offers solutions to increase building protection. Dr. Hamlet holds a DVM from Washington State University. He also holds veterinary licenses in Alaska, California, Massachusetts, Oregon, and Washington, as well as appointments at Tufts University, the Arctic Medical Research Laboratory, and as the sole civilian representative on an expert panel directed by the Secretary of the Army to review programs for high intensity training safety.

Stuart L. Knoop is co-founder of Oudens Knoop Knoop + Sachs Architects of Chevy Chase, Maryland. He has been involved in design for physical security for more than 30 years, particularly for the U.S. State Department, Overseas Building Operations, General Services Administration, National Institutes of Health,

Department of Veterans Affairs, and Walter Reed Army Medical Center. He has served on many NRC committees, including the Committee on Research for the Security of Future U.S. Embassy Buildings and the Committee for Oversight and Assessment of Blast-Effects and Related Research. He also served as vice chair of the Committee on Feasibility of Applying Blast-Mitigating Technologies and Design Methodologies from Military Facilities to Civilian Buildings and as chair of the Committee to Review the Security Design Criteria of the Interagency Security Committee. He is also a former member of the NRC Commission on Engineering and Technical Systems. Mr. Knoop is a registered architect, a fellow of the American Institute of Architects, and a member of the American Society for Industrial Security and the Construction Specifications Institute. He holds a B.Arch. from Carnegie Institute of Technology (Carnegie Mellon University) and was a Fulbright scholar to the Architectural Association in London, England.

Andrew Maier is the associate director for the nonprofit group Toxicology Excellence for Risk Assessment (TERA) and former manager of TERA's chemical risk assessment (VERA) program. He has led efforts at TERA in the area of occupational toxicology. While at TERA, he has coauthored technical reports, human health risk assessment documents, and toxicity summaries for environmental, consumer, occupational, or emergency exposure scenarios covering more than 100 individual substances. Development of these documents included critical review and analysis of animal toxicity, epidemiology, and mechanistic studies, with integration of this information for the derivation of human health risk values, including assessment of hazard and risk from all routes of exposure. Dr. Maier is active in risk assessment methodology research and has published in the areas of biomarkers, use of genetic polymorphism data in risk assessments, and methods in occupational toxicology. In addition to his Ph.D. (University of Cincinnati) and board certification in toxicology, Dr. Maier holds an M.S. in industrial hygiene (University of Michigan) and has been certified in comprehensive industrial hygiene practice since 1994. He has practical technical experience in occupational hygiene as a former industrial hygienist in private industry where he managed all aspects of a comprehensive industrial program, including hazard evaluation and control. In this capacity, he gained experience in evaluating exposure control methodologies for diverse industrial, research, and office facilities. He is an officer of the American Industrial Hygiene Association's Workplace Environmental Exposure Levels Committee and a member of the Society of Toxicology Occupational Health Specialty Section.

R. Paul Schaudies is the chief executive officer of GenArraytion, Inc., a small veteran-owned business that develops DNA- and RNA- based methods for identification and characterization of biological organisms. Before that, he was the assistant vice president and division manager at Science Applications International Corporation (SAIC). He was key in establishing the levels for reentry into the

Hart Building and is a nationally recognized expert in the fields of biological and chemical warfare defense. He has served on numerous national-level advisory panels for the Defense Intelligence Agency, Defense Advanced Research Projects Agency, and Department of Energy. He has 14 years of bench research experience managing laboratories at Walter Reed, the Walter Reed Army Institute of Research, and as a visiting scientist at the National Cancer Institute. He spent four years with the Defense Intelligence Agency as collections manager for biological and chemical defense technologies. As such, he initiated numerous intra-agency collaborations that resulted in accelerated product development in the area of biological warfare agent detection and identification. He served for 13 years on active duty with the Army Medical Service Corps and is a lieutenant colonel in the U.S. Army Reserve. Dr. Schaudies has served on many NRC committees, including the Committee to Review the National Nanotechnology Initiative.

Richard G. Sextro is director of the Indoor Environment Department at Lawrence Berkeley National Laboratory. He received his B.S. from Carnegie Institute of Technology (now Carnegie Mellon University) and his Ph.D. from the University of California, Berkeley. Dr. Sextro has been actively involved in research concerning biological and chemical warfare agents in indoor environments. He has recently completed a modeling study on indoor dispersion patterns of anthrax spores. Dr. Sextro served on the NRC Committee on Safe Buildings Program and Committee on Risk Assessment of Exposure to Radon in Drinking Water.

Linda D. Stetzenbach is a professor in the Department of Environmental and Occupational Health and graduate coordinator in the School of Public Health at the University of Nevada, Las Vegas. She received her B.S., M.S., and Ph.D. in microbiology from the University of Arizona. Her research interests are in the characterization of airborne and surface-associated microorganisms in indoor environments that affect human health; enhanced detection methodologies for identification and enumeration of microorganisms in environmental samples; fate and transport of airborne microorganisms (bioaersols). Dr. Stetzenbach served as an editor for the journal *Applied and Environmental Microbiology* from 2001 to 2004. She is serving on the American Society of Heating, Refrigerating and Air-Conditioning Engineers Special Project Committee (SPC 180P—Standard Practice for Inspection and Maintenance for HVAC Systems).

Linda M. Thomas-Mobley is an assistant professor in the College of Architecture's Building Construction Program at Georgia Institute of Technology. Dr. Thomas-Mobley has a Ph.D. from Georgia Institute of Technology; a J.D. from the University of Miami; and an M.S. and a B.S. in civil engineering from the University of Florida.

David R. Walt is the Robinson Professor of Chemistry at Tufts University and a Howard Hughes Medical Institute Professor. He is also the founding scientist, director, and chairman of the Scientific Advisory Board of Illumina, Inc. He received his B.S. in chemistry from the University of Michigan and Ph.D. in chemical biology from the State University of New York, Stony Brook. His laboratory is world renowned for its pioneering work that applies micro- and nanotechnology to urgent biological problems, such as the analysis of genetic variation and the behavior of single cells, as well as the practical application of arrays to the detection of explosives, chemical warfare agents, air contaminants, and food and waterborne pathogens. Dr. Walt received the National Science Foundation Special Creativity Award in 1995 and the 3M Research Creativity Award in 1989. He has served on a number of NRC committees, including the Committee on Review of Testing and Evaluation Methodology for Biological Point Detectors.

C

Presentations to the Committee

The National Academy of Sciences Building
Washington, D.C.
September 18, 2006

Perspective of Sponsoring Agency
Brian Reinhardt, Defense Threat Reduction Agency

Overview of the Immune Building Program of the Defense Advanced Research Projects Agency
Wayne Bryden, Defense Advanced Research Projects Agency

HVAC for Enhanced Building Security
Patrick Spahn

Report from Working Group on the Potential of Enhanced Building Filtration in Reduction of Anthrax Morbidity and Mortality Following a Bioterrorism Attack
Penny Hitchcock, University of Pittsburgh Medical Center

The National Academy of Sciences Building
Washington, D.C.
November 14–15, 2006

Immune Building Toolkit
Roger Gibbs, Special Projects Office

Overview of NIST Programs Related to Immune Buildings
Andrew Persily, National Institute of Standards and Technology

D

Acronyms and Abbreviations

AD	Aerodynamic diameter
AEGL	Acute exposure guideline level
AHU	Air-handling unit
ASHRAE	American Society of Heating, Refrigerating and Air-Conditioning Engineers, Inc.
ASTM	American Society for Testing and Methods
BPTK	Building Protection Toolkit
BSL	Biosafety level
BVAMP	Building Vulnerability Assessment and Mitigation Program
CB-EMIS	Chemical/biological emergency management information system
CCTV	Closed-circuit television
CONOPS	Concept of operations
COTS	Commercial off-the-shelf
CP	Collective protection
DARPA	Defense Advanced Research Projects Agency
DHHS	Department of Health and Human Services
DHS	Department of Homeland Security
DOD	Department of Defense
DOJ	Department of Justice
DOP	Dioctyl phthalate
DTRA	Defense Threat Reduction Agency

EO	Executive order
EPA	Environmental Protection Agency
FBE	Fraction of building exposed
FEMA	Federal Emergency Management Agency
FOE	Fraction of occupants exposed
FTIR	Fourier transform infrared
GAO	Government Accountability Office
GSA	General Services Administration
HEPA	High-efficiency particulate air
HVAC	Heating, ventilating, and air conditioning
IMS	Ion mobility spectrometry
ISC	Interagency Security Committee
IT	Information technology
JBPDS	Joint Biological Point Detection System
JOC	Joint Operation Center
LBNL	Lawrence Berkeley National Laboratory
LCAT	Life-cycle Cost Analysis Tool
LIDAR	Light-detection and ranging
LP	Level of protection
MERV	Minimum efficiency reporting value
MOS	Metal oxide sensor
NIOSH	National Institute for Occupational Safety and Health
NRC	National Research Council
OCC	Olympic Coordination Center
PCO	Photocatalytic oxidation
PCR	Polymerase chain reaction
PPE	Personal protective equipment
PROTECT	Program for Response Options and Technological Enhancements
TARA	Threat assessment and risk analysis
TIC	Toxic industrial chemical

TIM	Toxic industrial material
UFAD	Underfloor air distribution
UFC	Unified Facilities Criteria
UOPSC	Utah Olympic Public Safety Command
UV-LIF	Ultraviolet laser-induced fluorescence
UVGI	Ultraviolet germicidal irradiation
VFS	Ventilation and filtration air-handling systems
WMATA	Washington Metropolitan Area Transit Authority

E

Glossary

Aerodynamic diameter (AD); aerodynamic particle size	The equivalent spherical diameter that approximates the aerodynamic behavior of an irregular-shaped particle.
Air-handling unit	A component of a heating, ventilating, and air-conditioning (HVAC) system that delivers conditioned air and outside air to spaces in a building.
Arrestance test	A filter efficiency test according to the American Society of Heating, Refrigerating, and Air-Conditioning Engineers (ASHRAE) Standard 52.1 (ASHRAE, 1992) that utilizes an ASHRAE synthetic dust and the mass fraction of dust removed is determined.
Asymmetric warfare	A military situation in which two belligerents of unequal strength interact and take advantage of their respective strengths and weaknesses.
Atmospheric dust-spot efficiency	An evaluation used for high-efficiency filters, defined by ASHRAE Standard 52.1 (ASHRAE, 1992), that involves the use of unconditioned atmospheric air passed through the test material, and visualization of discoloration of the air downstream is compared to upstream (unfiltered) air.

Billeting	Any building or portion of a building, regardless of population density, in which 11 or more unaccompanied personnel are routinely housed, including temporary lodging facilities and military family housing permanently converted to unaccompanied housing.
Bioaerosol	A collection of airborne biological material including virus, bacterial cells, fungal spores, fragments, and by-products of microbial metabolism.
Building envelope	The entire enclosing construction for a building, including walls, windows, doors and other openings, and roof.
Clinical sign (of disease)	Objective evidence of disease especially as observed and interpreted by a physician rather than by the patient or a lay observer.
Collective protection	A provision of a contaminant-free area in which personnel can function without individual protective equipment, such as a mask and protective garments.
Compartmentalization	A division of interior spaces into separate, discrete spaces (see also "Zoning").
Cost-benefit analysis	A technique for project appraisal that weighs the total expected costs against the total expected benefits of one or more actions in order to choose the best or most profitable option.
Cyclone collection or separation	A method that uses vortex flow to separate different sized particles.
Detect to treat	Provide information in the time frame needed to initiate treatment options to minimize adverse health effects from exposure.
Detect to warn	Provide information to initiate a response action that minimizes exposure.

Detection system	A system that can recognize the presence of classes of threat agents but generally does not include identification. When unqualified, detection includes all the possibilities shown in Figure 3-1: clinical, symptomatic, visual, technology-based signatures, and assays. A detector technology system would refer to a technology-based detection system.
Detection technology	Technology-based detection.
Detector	A device that generally provides information on the presence of a threat agent or class of agent but does not reveal identity and is usually continuously operated. (See Figure 3-1 for types of detectors.)
Ducted return	The use of an air-handling duct system to return air from the interior of a building to the air-handling unit.
Dust-holding capacity	The amount of dust that an air cleaner can retain when it is operated at a specific airflow rate to a maximum resistance value.
Dust-spot filter	A filter that is reported to remove 85 percent of particles 2.5 μm in diameter.
Electrostatic filtration	The use of a charge field in a filtration unit to remove charged particles. Particles could be naturally charged or given a charge prior to filtration.
False alarm	A wrong indication that a threat agent is present.
Filter bypass	The airflow around the filter matrix caused by channeling of the filter material, overloading of the filter, use of incorrectly sized filters, or improper installation.
Filter efficiency	The ability of a filter to remove particles from the air-stream.
Filtration	A collection of particles larger than the filter pore size.
Heating, ventilating, and air-conditioning system	A system designed to provide conditioned air to the various spaces in a building in order to satisfy demands for heating, cooling, humidification, dehumidification, and contaminant removal.
High-regret options	Options to respond to a situation that could incur a high degree of remorse or serious consequences.

Identification system	A system that determines the specific threat agent such that complete response actions, including medical treatment, are possible. (A caveat is that for unknown threat agents that have emerged or are engineered, identification can be delayed, such as for severe acute respiratory syndrome.) These can be exclusively technology based or human based or both.
Impactors	A collection system in which particles are accelerated in a jet toward a surface (classical impactors) or toward a nozzle (virtual impactors).
Impingement filtration	The retention of a particle when it is either too large or dense to follow the airstream around a fiber so that it lands on the surface and is retained due to attraction.
Infectious dose	A number of pathogenic microorganisms needed to cause disease.
Infectivity	The ability of pathogenic microorganisms to infect (that is, enter and multiply in) the cells of a host's body.
Infiltration	A flow of air from the exterior of a building to the interior.
Inhabited building	A building or portion of a building routinely occupied by 11 or more personnel and with a population density of greater than one person per 40 gross square meters (430 gross square feet).
Leakage	The flow of air through the building envelope—may be from the outside of a building to the interior or from the interior to the exterior.
Life-cycle cost	All possible expenses incurred during the lifetime of a system. Life-cycle cost includes initial costs of implementation and all expenses incurred during the period of operation of equipment or a system.
Low-regret options	Options to respond to a situation that could incur little remorse or not-so-serious consequences.
Metal oxide sensor	A sensor that responds to virtually all organic vapors and provides information that a vapor release has occurred with little or no identification capability.

Minimum efficiency reporting value (MERV)	A filter rating, defined by ASHRAE Standard 52.2 (ASHRAE, 1992), that is based on filtration efficiency as a function of particle size. Higher MERV values indicate more efficient filters.
Natural ventilation	The use of open windows to supply outdoor air to interior spaces of a building.
Outside air intake	A point of entry of outside air delivered to the spaces in a building by an air-handling unit.
Particulate	Airborne solid material.
Pathogenicity	The ability to cause disease.
Plenum return	Unducted spaces, typically the cavity between the suspended ceiling and the floor or roof above, in which air from multiple areas in the building mixes before returning to the air-handling unit or being evacuated to the outside.
Polydispersed aerosols	Aerosols whose particles are of various sizes.
Polymerase chain reaction (PCR)	A molecular biology amplification method to detect and identify DNA sequences.
Prefilters	Large porosity filters used to remove large particulate debris prior to the use of smaller porosity filters.
Remote sensing	The ability to have a sensor at one location and to detect and identify the presence of a particular object at another.
Sensor network for distributed sensing	The spatial distribution of sensors.
Sensor system	A technology-based detection and identification system of threat agents that are localized at the collection point (for example, Biowatch is not a sensor system). Although sensor systems have the goal to identify the threat agent, this would be possible in many current applications only for limited sets of threat agents or would require off-site confirmatory identification.

Sensor technology	A technology that provides information on the presence and possibly the identification of a threat agent or class of threat agents. Sensors and sensor technology may be operated intermittently or continuously.
Standoff	The distance between a target (e.g., building) and a potential hostile event (e.g., explosion or chemical or biological agent release point).
Supply ductwork	A positively pressurized conduit, typically made of sheet metal, used to distribute supply air to air delivered from the air-handling unit to spaces within a building.
TaqMan PCR	A commercially available PCR method for detection of DNA and RNA.
Test bed	An environment created for testing that contains the integral hardware, instrumentation, simulators, software tools, and other support elements to approximate a real-world situation.
Threat agent	An agent used to inflict damage to a facility or harm to its occupants. Threat agents include biological, chemical, and radiological agents.
Threat type	The combination of a threat agent and an implementation strategy used to inflict damage to a facility or to harm its occupants.
Tiered detection systems	The staged deployment of typically inexpensive, fast-acting, low-accuracy detectors followed by more accurate sensors or confirmatory testing. Tiered detection systems are typically used to minimize the need for costly confirmatory tests. Tiered detection can also be spatially dispersed.
Transmissible	The ability of pathogenic organisms to be transmitted from one person to another.
Trigger	A detection technology that initiates an action and is generally more rapid, lower cost, and less specific than the identification technology.
Tropism	An involuntary response of an organism or part of an organism toward or away from external stimuli.

Ultraviolet germicidal irradiation	The use of various wavelengths of ultraviolet light (UVC, 200-280 nm and UVB, 280-315 nm) to inactivate biological materials.
Underfloor air distribution	A system that utilizes underfloor space for distribution of air rather than overhead plenum space or supply ductwork.
Uninhabited buildings	Spaces not considered inhabited, primary gathering, or billeting.
Zoning	The process of defining areas served by independent HVAC systems.